Intrahousehold resource allocation and well-being

Intrahousehold resource allocation and well-being

The case of rural households in Senegal

AWLAE SERIES No. 10

Fatimata Dia Sow

Wageningen Academic
Publishers

ISBN: 978-90-8686-158-3
ISBN: 978-90-8686-714-1
DOI: 10.3921/978-90-8686-714-1

ISSN: 1574-2474

First published, 2010

Wageningen Academic Publishers
P.O. Box 220
6700 AE Wageningen
the Netherlands
www.WageningenAcademic.com
copyright@WageningenAcademic.com

This thesis is dedicated:

to my mother Ramatoulaye Watt
(lady of piety and good heart)

to my late brother Alioune Dia who left us so soon

to my husband Oumar Sow and my children
Ndeye Sokhna, Sokhna Mame Diarra and Sokhna Bineta

to all women of good will

to my country Senegal

Acknowledgments

Thanks to the Almighty and faithful God who gave me life, health and strength to successfully achieve my PhD thesis. He is the one you give to whom he wants, when he wants.

When I started writing this PhD thesis almost five years ago, my challenge was to bring a valuable contribution to the knowledge in the economic field on gender issues in my own country. Although, these five years were an episode of hope and many sacrifices because of the hard and tough way that this challenge implied. Coming from a French speaking country, the first challenge was to work in English which at its own was a great part of the Phd success. On several occasions, I felt unable to do because of the many pitfalls and difficulties. So far, I may note that because of many encouragements and revisions, the finished product made me reasonably happy and satisfied. I am exceedingly grateful and indebted to several persons and institutions.

My sincere thanks go to the Government of the Netherlands who funded this scholarship in collaboration with Winrock International through the African Women Leaders in Agriculture and Environment (AWLAE). I am also indebted to my mother institution ISRA (Senegalese Agricultural Research Institute) who allowed me to get so high in terms of training and knowledge. I thank the General Directors who have succeeded at the head of ISRA during this long period of training: Dr. Pape Abdoulaye Seck, Dr. Taib Diouf and Dr. Macoumba and the scientific staff, Dr Alioune Fall and Dr Abdoulaye Fall. My thanks also go to my colleagues in the Animal Production Department (LNERV) and to the administrative staff of ISRA: Dr. Thiongane, who particularly facilitated the field work, Ms Khoudia Bass, Assy, Leo and all those who supported me and encouraged me during this long period of hard work. My thanks also go to the African Economic Consortium (AERC) for having supported part of my extension program.

I do not have enough words to thank and congratulate my superviser Prof. dr. Gerrit Antonides who spared no effort to support me morally and scientifically. As a great scientist, you believed in my skills and step by step contributed to the success of my proposal and the development of this thesis. Nothing in this thesis has been carried out without your caring, your tireless review of the different versions of the chapters, your wisdom and positive guidance and criticism. The first quality of a true and great scientist is to achieve a scientific construct and outcome from difficulties and challenges. Of course, you are a great scientist and I would always be grateful for your involvement in the reflection and achievement of this work. To me, you are an example of excellence and competence. I also really appreciated your visit in Senegal during the field work where you learnt to understand the realities of the field research and gave great insights in the implementation of the database. I also truly appreciated the company of Melanie and Rianne in the field and our visit to our religious city, Touba.

My thanks and gratitude also go to Prof. dr. Anke Niehof who is an examplary woman for her running battle within the AWLAE program where each of the theses of the 20 beneficiaries of the AWLAE program is a challenge for her personal success. You put much effort and sacrifice into the completion of the AWLAE program. For us who have benefited from this program, you are a leader and guide in the promotion of African women at high level as intellectual and leaders in the world. Beside the late Julia Jitobu and also Lisa Price, you have shown us the way through the challenges of today and tomorrow. As a supervisor, you put all your patience and technical expertise for the success of this thesis. I also appreciate the profound contributions of Dr. Johan van Ophem, my co-promoter, whose advice on several occasions during the discussions gave more insight in our work and strengthened our major ideas. Thank you also for the Dutch summary.

I extend my thanks to Dr. Pierre van Mouche to whom I am highly indebted. I really appreciate your contribution and insights in the modeling of some chapters of my PhD thesis. You put a large part of your time into working with me as an adviser; I appreciate your commitment and willingness to have accompanied me in the mathematic field, thank you for the high quality work you brought into my thesis. My thanks go also to Prof. dr. Wim Heijman for the excellent moment we shared at the ECH department.

This work would not have been possible without the commitment of my numerators who were deeply engaged in the field work and have achieved a high quality work. My thanks go to Souleymane Ndiaye, Mamadou Ba, Djiby Ka Boyli Ka, Sene, Amadou Demba Sall and to their family.

I cannot fail to thank Fat Cheick, you were more than a friend but a sister who helped and assisted me in many solicitations. Thanks you for your support in the field and for your involvement in the investigations. I am also indebted to your husband Sane during our long stay at Dahra. Thank you.

My profound gratitude is also due to the support of the management team of the AWLAE scholars PHD program who were essential in facilitating administrative matters of the research work: Nora Muthike, Hedy Munro, Margaret van Wissen and Annelies Coppelmans. Thank you Margaret for the hard work you have done and all the time spent in editing the document. Without your understanding and support, it would not have been easy to finish in time. You shared administrative and financial difficulties with us and showed patience and commitment for ensuring a successful completion of the AWLAE program.

I thank my colleagues of the AWLAE Program whom I would like to call the Iron Ladies, for their involvement and total dedication and unwavering respect for their noble commitment. I would mention particularly Aifa Ndoye with whom this all started and with whom I have come a long way through sacrifice, tears, heartbreak, but ultimately success built on a deep friendship and infallible confidence. Beyond the AWLAE program, I am sure that we will share other exciting programs in our own country and throughout the world: thank you. Thank you also my sisters Mariame Maiga and Fofana Binate. Mariame, I will be always indebted for your generosity and your kindness. You will always be my jovial Mariame who usually made me laugh when I was so stressed. With Namizata I shared unforgettable moments of meetings, discussions and exchanges, thank you for the sisterhood. You will always be part of my family as sisters because of 5 long years of journey through sharing joy and sadness. I am also grateful for the friendships forged with Lydia, Hirut, Stephanie, Faith, Rose, Susanna, Joyce, Ekaete, Corrie, Carolyne, Doris, Gaynor, Kidist. Thank you for your encouragement and patience. I will never forget your visit to Brussels when Bineta was born. This is part of my best memories during the program. Also, I cannot fail to thank my colleagues of the Economic of Consumer and Household Group: Romanus Dimoso, Leonie Cramer and Marleen Schouten.

I am forever indebted to my family who kept me going without resigning, supporting me morally and spiritually. In the first place my darling husband Sow Oumar who has contributed to all stages of success in this effort through moral support and assistance. You have endured my many absences and you never complain or give up, helping me going beyond my abilities, praying and supporting a large share of my failures. Thank you for everything. My thoughts also go to all your family, Sokhna Bineta and the children, Mame Seydou, Balla, my dear Fati...

My thoughts go to my babies Ndeye Sokhna, Mame Diarra and Bineta who suffered upheavals in their own childhood because of my repeated absences and have learned to grow quickly. You have been my moral support. This work is also the fruit of your patience and understanding despite your young age. Be always united and work well to go beyond that success in your life. You could not get by without the help and support of my sister Mariame Dia and her husband Abou Ly, who have devoted all their efforts in your daily care and help. Thank you. My lovely tought goes to Baby Allassane Aliou.

Thank you too my parents who prayed unceasingly and to whom I wish long life and health, Baba, my mother, Gorgol Awa Ly, Baba Cité and his family, the late Alioune Dia, my uncles and aunts. My thanks also go to all my brothers, my cousins, my sisters, nephews and nieces. I extend these thanks to Suzanne Dia who assisted me at the earlier stage of my writing proposal. To my oldest sisters who learned to replace me at home during all our family events, thanks you and be united forever: Ramatoulaye Dia, Awa Dia, Ndeye Aidara, Thilo, Sala, Aissata and Defa Dia. All my thanks to your husbands too: Djiby Watt, Djiby Diagne, Djigo, Baro Ndiaye and Pape Ndiaye. My thoughts also go to my brothers Abou Dia, Oumar, Ousmane, Seydou Nourou.

To my close friends who also believed in, and supported me throughout the whole process, I express my sincere appreciation of your assistance. Thanks you especially Adama Faye, Ndèye Mbodji, Amy Badiane, Safiétou Fall, Astou Sène, Atte, Charlotte Diop, Catherine, I greatly appreciate your encouragements and concern.

I cannot forget to thank my friends from Belgium and Wageningen, Oumy Khairy Ndiaye from CTA, Allassane and his wife Maimouna Mbodji, Kadidiatou Seck, Josephine, Amadou Ba and his family at Nijmegen and all the others that I cannot name for their support and prayers.

Last but not the least, I extend my gratitude and thanks to Serigne Mame Balla Mbacke who stood with me in prayer during all the steps of this PhD thesis. I will always be indebted to you for your support and prayers.

Wageningen, the city of life sciences

Table of contents

List of tables

List of figures

List of acronyms

ANCAR	National Agency for Rural Agricultural Advice (Agence Nationale Pour le Conseil Agricole et Rural)
ANDS	Senegalese Agency of Statistics and Demography in Senegal (Agence nationale de la statistique et de la démographie du Sénégal)
AWLAE	African Women Leadership in Agriculture and Environment
BAD	Development Bank of Africa (Banque Africaine de Développement)
CAPEC	Popular Agency for Credit and Savings (Caisse Populaire d'Epargne et de Crédit)
CARITAS	Confederation of Catholic Care Organizations
CMS	Mutual Agency of Senegal (Caisse Mutuelle du Sénégal)
CNCAS	Senegalese Agricultural Bank of Credit (Caisse Nationale de Crédit Agricole)
DIRFEL	Directory of Women in Livestock (Directoire des Femmes en Elevage)
DJOMEC	Mutual Agency for Credit and Savings of Djolof (Mutuelle d'Epargne et de Crédit du Djolof)
FAO	Food and Agriculture Organization of the Unites Nations
FONGS	Senegalese Federation of Non-Governmental Organization (Féderations des Organizations non Gouvernementale du Sénégal)
GDP	Gross Domestic Product
GUANA	Big Agricultural Offensive for Food and Abundance (Grande Offensive Pour l'Alimentation et l'Agriculture)
IFAD	International Fund for Agricultural Development
IFPRI	International Food Policy Research Institute
ILO	International Labour Office
IMF	International Monetary Fund
ISRA	Senegalese Agricultural Research Institute (Institut Sénégalais de Recherches Agricoles)
MDE	Herders' House (Maison des Eleveurs)
MEAH	Micro-Ecological Approach to Health
MEF	Ministry of Economics and Finance (Ministère de l'Economie et des Finances)
NGO	Non-Governmental Organization
PAMECAS	Senegalese Mutual Agency Partnership for Saving and Credit (Partenariat des Mutuelles d'Epargne et de Crédit du Sénégal)
PAPEL	Livestock Assistance Project (Projet d'Appui a l'Elevage)
POGV	Village Organization Program (Projet d'Organization Villageoise)
PMIA	Agricultural Modernization and Intensification Project (Projet de Modernisation et d'Intensification Agricole)
PNDS	The National Plan for Health Development (Plan National de Développement de la santé)

PNUD	United Nations Development Programme
PROMER	Project for the Promotion of Rural Micro-Entreprises (Le Projet de Promotion des Micro-Entreprises Rurales)
PRSD	The Poverty Reduction Strategic Document (Document Stratégique de Réduction de la Pauvreté)
PRSP	Poverty Reduction Strategic Paper in Sénégal (Document Stratégique de Réduction de la Pauvreté)
REVA	Plan to Return to Agriculture in Senegal (Plan de Retour au Senegal)
SODEFITEX	The Enterprise of Development of Textile Fiber (Société de dévelopment des Fibres Textiles)
SPSS	Statistical Package for the Social Sciences
STATA	Data Analysis and Statistical Software
UNAIDS	Unites Nations on AIDS
UPPRAL	The Union of Milk Producers in the Pastoral Zone at Dahra (l'Union des Producteurs et Préposés du Rayon Laitier de Dahra)
WHO	World Health Organization

Glossary

dioula	intermediary traders in livestock (buying and selling livestock for owners)
tefanke	are retailers selling livestock products
soudure	dry period from March to April-June
Diom Galle	Head of Household in the Peulh language
Borom Ndjel	Head of Household in the Wolof language

Chapter 1

Introduction and research problem

1.1 Context

The Senegalese economy, like that in many West African countries, faces multiple socio-economic constraints that handicap the local economy and development process. At the macroeconomic level, the World Bank's Structural Adjustment Program (SAP) was implemented early in 1980 to promote economic stabilization and structural changes, and to improve foreign investment and reduce government deficits. However, SAP largely failed to meet expectations. In the 1990s, twenty years after the first SAP, Senegal fell into debt traps which increased poverty and several disparities (wealth, welfare, income) between farmers and the urban population (Durufle, 1994; Diouf, 2002). For small farmers, the situation got worse due to the marginalization of the rural poor, especially women and children. It is these who were most affected by the SAP. The marginalization of the poor was due to the abolition of subsidies for inputs and having no access to formal credit for lack of collateral. In fact, SAP increased gender inequality by neglecting the fact that all economic activity works through and within gendered relationships (Elson, 1993). Moreover, SAP and the devaluation of the Senegalese currency in 1995 contributed to a reduction of employment, paid work (Rouis, 1994; Lee and Loufti, 1992) and income: minus 0.3% during 1960-1978 and minus 0.6% after SAP between 1985 and 1990 (Ndiaye, 2001; MEF, 1997). This situation mostly led to a decline of key sectors like agriculture, education and health, and contributed to a delay in development of infrastructure. At the micro level, this situation induced a dislocation of couples and the disintegration of families, which was recently amplified by the HIV/AIDS epidemic as part of the health problem (Mutangadura, 2000; Müller, 2004, 2005; UNAIDS, 2003, 2006, 2008). In fact, mortality in West-Africa including Senegal, is the highest in the continent (WHO, 2003). Recently, this situation contributed to a reduction of the life expectancy: now 49 years in West-Africa as compared to 60 years in East Africa and 65 years in North-Africa (WHO, 2003).

The Senegalese economy depends mostly on the agricultural sector. Despite the total economic growth which started in 1996 (4.2%) thanks to the growing importance of activities related to fishing and services, the agricultural sector remained stationary due to the lagging production of cash crops. Major examples (peanut and cotton) remained vulnerable to international market threats and local trends in the socio-cultural and natural environment (drought). Moreover, the decline of average rainfall reduced the production of crops, especially of peanuts and millet, and the livestock and income of farmers. As a consequence, Senegal is one of the 10 poorest countries in the world (IMF, 2007; Annabi *et al.*, 2005). Thirty-eight percent of the people at the national level are poor, and 67% of these live in rural areas (Senegal/MEF, 1997).

The reforms initiated within the framework of the 2002 Poverty Reduction Strategic Paper (PRSP) (IMF, 2006), have among other priorities, promoted agriculture and gender equality and women's economic independence by aiming at improving their economic status (access to credit, better education, funding women's development projects). The substantial shares of the total budget allocated to agriculture (25%), and to health (19%) did not favour the primary sector's growth which remains low, i.e. 2% in 2004 and below the 7-8% needed to reduce poverty by half before 2015. In fact, those living in poverty comprise more than half of the population, although poverty has slowly decreased over the period 1994-2004 (68% in 1994 to 54% in 2004). Nowadays, rural Senegalese household income is primarily generated by non-agricultural activities (28%), salaries (24%) and money transfers (19%) (Fall, 1997; Fall *et al.*, 2004; Gaye *et al.*, 2004). Moreover, studies show that 50% of rural households are not self-sufficient in the supply of cereals (Adjamagbo *et al.*, 2006). Between 1994 and 1997 the share of agricultural income in total income declined by 11% of GNP (see Ndiaye, 2005).

1.2 Problem statement

In Senegal, production types vary across sylvo-pastoral (mainly livestock) and agro-pastoral (livestock combined with cash crops and food crops) production systems, the availability of land being one of the critical factors. In the past few years, food shortage became pronounced and extensive, especially during the annual hunger season or 'soudure' from March to September. Policies implemented by the government to overcome poverty through support, extension programs and innovative technologies, generally remain insufficient. Projects implemented among male farmers to improve the production of cash crops, farm organization and equipment, did neither assist nor improve the capabilities of rural women. Moreover, most policies implemented lack attention to improving the productivity of rural women. Because of the neglect of gender differences in policy decision-making, women are put at a disadvantage by its instruments and interventions implemented. Programs and projects have focused mainly on the male head of the household, as the right target to take care of the other household members, thus under-valuing the intrahousehold gender relations that maintain the survival of the household (Elson, 1991). Even though substantial effort taken in the last few years to improve women's living conditions in rural areas (e.g. distribution of small equipment aimed at reducing the work burden, the duration of the micro credit), little has been done to analyze sex roles and gender responsibilities and to use the analysis in policy adaptations. Despite the fact that women in many cases have been working hard in the farming system, their lack of voice and of access to resources (land) are recognized as the main constraints to their participation in the development process (Mehta, 1998; Kelly *et al.*, 2002).

Advanced research into crop growing and women's roles in developing countries showed the micro- and macroeconomic impact of production, especially of cash crops, on the gender division of labour and income (Elson, 1991; Mbilinyi, 1988; Lado, 1992; Darity, 1995). Following this line of research, other research in sub-Saharan Africa resulted not only in a better understanding of household responses to gender-differentiated priorities under specific policy (such as price policy), but also in the necessity to incorporate a gender approach in analysing and designing policy instruments (Warner *et al.,* 2000). The gender inequality in the implementation of policies induced, among other consequences, lower productivities and favoured low outcomes in farming systems (Warner *et al.,* 2000; Lado, 1992).

In the last two decades, a body of knowledge has been developed regarding intrahousehold interdependencies in farming and the role of women in food security (Fafchamps, 1998; Haddad *et al.,* 1997; Niehof, 2003). The relevance of understanding the separate roles and responsibilities of men and women lies in the importance of understanding the allocation of resources (e.g. capital and labour) and the position of each actor in the household. For instance, the interdependence between men and women occurs with respect to (1) the labour and time allocation to agricultural activities and (2) the decision-making about the allocation of resources and the consumption pattern. These two aspects have an impact on their welfare and their well-being and that of the household as a whole. The allocation of resources determines the management of agricultural and non-agricultural activities and hence the household earnings. Since labour and time allocation of men and women differ because of specialisation in specific activities, it would be relevant to understand how labour and time of household members affect household production and earnings. The intrahousehold division of labour and time may reflect differences in preferences possibly shaped by differences in local norms, socio-economic factors and intrahousehold bargaining (Fafchamps *et al.,* 2005). The distribution of time by gender responds to economic incentives and constraints (Ilahi, 2000). For example, the burden of housework at the level of the household (in providing basic services as water and wood used for energy) reduces the involvement of women in home production and in earnings. A policy reform that nowadays increases women's empowerment by improving their access to basic services and means (water, credit, opportunities to involve in off-farm activities) has probably an impact on the female resource allocation and hence affects household production.

Moreover, differences in resource allocation may explain in large part the consumption pattern of the household members (Haddad *et al.*, 1997, World Bank, 2005). Because consumption depends on self production and expenditures, policies aiming at improving it should collect information about social and economic factors affecting consumption. One important issue is the interaction that takes place between husband and wife in decision-making regarding the consumption pattern. This issue is little understood in most countries because of the misunderstanding of the gender aspect in household decision-making (obligations, priorities, constraints). For this reason, it will be relevant to study and to contribute to understanding the relationships between men and women in households, and their associations with household expenditures and welfare and well-being. For instance, because priorities, obligations and access to resources differ between men and women, the control over income and the interactions that take place between men and women in the management of activities, greatly influence decision-making and responsibilities in the household but also efficiency of policy instruments. To the extent that individual earnings affect the bargaining and decision-making over consumption of goods (food, health, schooling and others) within the household, some important aspects (local norms, cultural believes, level of education, women's power in decision-making, etc.) may have great consequences for intrahousehold welfare. For example, it has been argued that an increase of women's income in the household leads to an improvement of livelihood conditions (education, food security and health of household members (Haddad *et al.,* 1997; Rosenzweig *et al.,* 1989; Lundberg, 2005; Browning *et al.,* 1994). Hence the individual power and preferences of household members impact on the welfare and the well-being of the household as a whole. This issue is the focus of our interest.

The aim of this study is to contribute to the idea that agricultural policies should be designed according to gender differences in preferences and resource allocation (Farrington *et al.,* 2003). The study is part of the African Women Leadership in Agriculture and Environment (AWLAE) project which focuses on two major problems that households encounter in developing countries: gender-biased intrahousehold dynamics and HIV/AIDS as part of health issues. In fact, the lack of consideration of gender issues greatly contributed to inefficiencies in agricultural activities and to food insecurity (Grown *et al.,* 2000, Hands, 2001; World Bank, 1997).

Health status which is part of our concern constitutes of course an important factor that can shape production and income and may not be favorable to women in terms of bargaining position and ability to face poverty and to contribute to food provision. Because of social (gender) and biological (sex) differences, men and women face different health problems (WHO, 2006a and 2006b). In the past decade such shocks have contributed to the extent and increase of poverty and vulnerability. Also policies have focused to improve health status because of the important role of health status in economic productivity. These policies have especially emphasized women and children because of their special needs. Despite the remarkable progress assessed through health indicators like mortality and morbidity rate in some countries, the health issue remains an important cause of low income and low productivity (WHO, 2003).

It has been recognized that the health issue has an important linkage with the gender issue and poverty. In fact the gender issue is critical because of the decision-making regarding access to health care (women are mainly responsible for care) and health facilities and socio-cultural constraints. The socio-cultural constraints are related to practices and beliefs of communities or preventive attitudes in case of illness, especially chronic illness (Chafetz, 2006). Evidence from many countries (Latin-America, Tanzania, Gambia) has shown that ill-health, especially chronic illness, causes variability in the efficiency of gendered production and household tasks, and in welfare and well-being across households (Browner, 1989; Lukmanji, 1992; Tapouzis, 2000; Rugalema, 1999a). For example, in the case of HIV/AIDS, several authors have contributed to highlighting the impact of ill health on the livelihood of the household in Sub-Saharan countries (Rugalema, 1999b; Mutangadura, 2000; Müller, 2004; 2005). In many African societies, the inequity in resource allocation (mainly disfavored to women) and decision-making regarding resource allocation and human capital endowment limit vulnerable people (women, children) to access health care and health facilities. The mis-understanding of these important determinants has led to policy failure in implementing projects and programs aiming at better health status. Despite the relatively low prevalence of HIV/AIDS in Senegal (less than 2%), it may still represent a threat to productive labour (UNAIDS, 2003). Indeed, the fact that women are physically more vulnerable to HIV infection and other health problems than men makes them also more vulnerable to the loss of control over resources and economic benefits (FAO, 2003, 2002). This vulnerability can have a direct linkage with their time allocation and with their earnings. For example, in most developing countries like Senegal, differences in health status between males and females have induced socio-cultural and economic effects on, for example, the role of women as caregivers, the gender specialization within household tasks, access to the labor market and the role of institutions (availability of infrastructure like roads, health centers or hospitals, access to medication, access to education, and traditional or modern knowledge). For example, relevant studies on gender and resource allocation at the household level have emphasized the important role of decision-making in the allocation of time to health care. Time allocation to health care of household members may likely impact on the allocation of labor in agricultural activities and earnings. Therefore, it is expected that time devoted to health care mostly done by women may be associated with household and individual earnings. This issue is the focus of our interest in understanding how ill health can be related to intrahousehold allocation of resources, division of labour, earnings and decision-making regarding expenditures.

In Senegal, farming activities are gendered in different ways because the gender division of labor, resources and responsibilities differ with respect to the type of farming system prevailing in particular areas. In this study, the choice of two research areas, agro-pastoral and sylvo-pastoral, with different systems of agricultural production will highlight the gender-based intrahousehold resource allocation and decision-making in a different context of production. The gender-based approach of this research leads then, to several research questions, see Section 1.4. This study seeks to provide more insight into the household allocation of resources on production and ill-health. Furthermore, it analyses the decision-making power of husbands and wives in the household regarding the responsibility for particular expenditures and the welfare and well-being of spouses in rural farming systems in Senegal.

1.3 Research objectives

The problem addressed in this research is the gender-based intrahousehold heterogeneity in household-resource allocation and its consequences for the income and welfare status of men and women in rural Senegal.

The main objective of this study is to find factors that contribute to enhancing welfare, bringing out poverty alleviation and increased earnings for smallholder producers in Senegal. Specifically, the study intends:

1. to gain insight into the factors affecting gender segregation in farming and household tasks and to assess the impact of resource allocation (labour, time, land) on access to production and income;
2. to gain insight into the way in which husbands and wives make decisions about daily consumption of goods and the consequences for woman empowerment;
3. to investigate how households with different patterns of income distribution cope with expenditures, especially with respect to resource allocation;
4. to examine the bargaining relationship between husband and wife (and its consequences on expenditures), and their welfare and well-being.

1.4 Research questions

1. How does the distribution of intrahousehold activities, decision-making and responsibilities in farming differ across sylvo-pastoral and agro-pastoral areas?
2. What are the structural and socio-economic dynamics and determinants that are associated with the gender division of resource allocation and control over income across different systems of production?
3. How does the gender allocation of labour and time devoted to agricultural activities, housework and health within the household may be correlated to the earnings of agricultural activities?
4. How do men and women engage in food and health provision in terms of decision-making, and how can we measure women's power?
5. How can we capture the bargaining relationship between husband and wife, and its effects on expenditures, and how can we measure their welfare and well-being?

1.5 Study design and approaches

We attempt to answer these questions within the analytical framework in Figure 1.1 using four different analyses in two different farming systems applied in the sylvo-pastoral area and the agro-pastoral area, respectively:

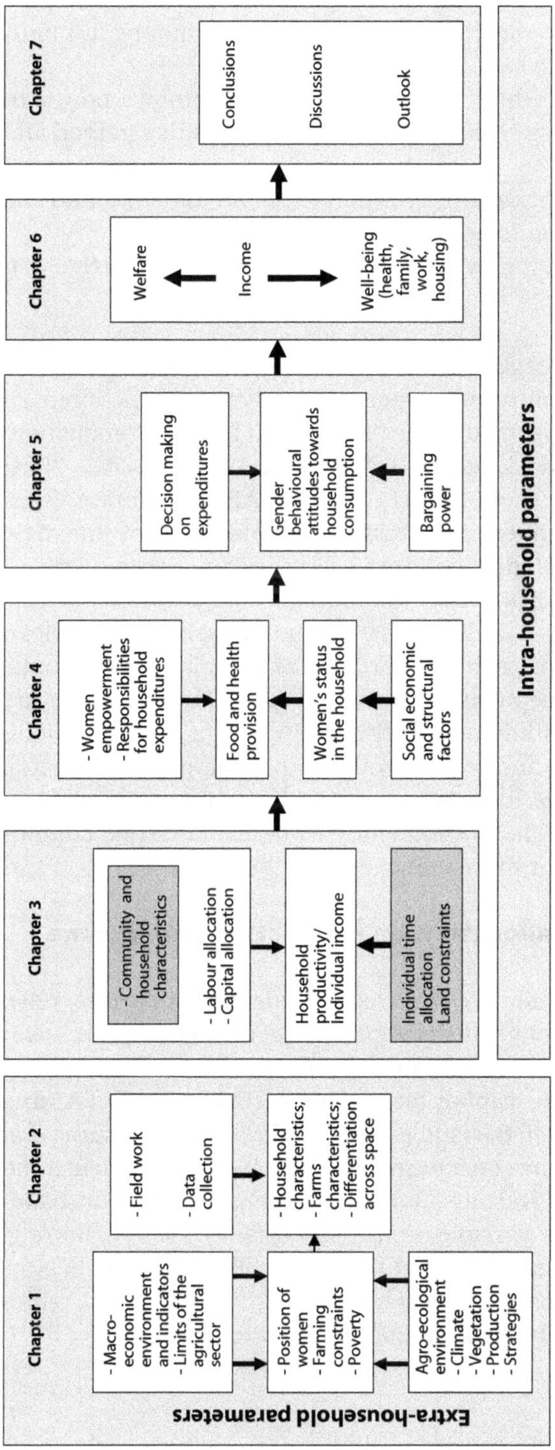

Figure 1.1. Analytical framework and overview of the thesis chapters.

1. the analysis of the determinants of earnings and time allocation among different agricultural activities (crops and livestock farming);
2. the gender-perspective approach which intends to analyze women's power in the household and decision-making regarding tasks, responsibilities in food and health provision;
3. the analysis of the bargaining relationship between husband and wife and the gender- behavioural attitudes towards household consumption;
4. the analysis of welfare and well-being by using a combination of subjective and objective approaches.

The analysis of the socio-economic organization of farming and gender relations requires a basic understanding of community structure and dynamics. Economists and sociologists seem to prefer the household as the basic unit of farming, production, consumption and decision-making (Mackintosh, 1989; Kabeer, 1994; Fall *et al.*, 2004; Niehof and Price, 2001; Mtshali, 2002; Niehof, 2004). In many African communities, the household is synonymous with a sphere or co-residence represented by one male head or extended to several families (adult children's families and other relatives' families) under the supervision of a 'patriarch,' the family being defined in terms of marital and parental relationships (Gardiner, 1997). The household, in African societies at least consisting of one couple, can be referred to as the sphere of economic production and consumption, income accumulation and survival (Antoine *et al.*, 1995; Fall *et al.*, 2004). The present study adopts this definition of the household. The household is a supplier of labour and other resources (capital, physical assets). Household members have productive functions, especially the provision of food, livelihood and care (Gardiner, 1997). The present study focuses on married couples with dependent children and/or other members (relatives, helpers).

1.5.1 Effects of labour and time allocation on production and income

In the analysis of labour and time allocation we shall apply gender theory (Ferree, 1990; Potuchek, 1992; Overbeek *et al.*, 1998) or gender-perspective theory (Thompson, 1993) to gain insight into the gender organization of tasks and time use. We aim at depicting the effects of labour time and human capital on gender participation in farming associated with different activities of both the landless and landowners. We assume that farm production and housework depends on a number of variables including income from different activities (crops, milk, livestock, homework) of individuals (husband, wife), labour time devoted to activities, variable inputs (seeds, feed), fixed inputs (e.g. land) associated with activity, human capital of individuals (skills, education, health) and other household characteristics. In this approach, we desegregate all the variables related to gender and estimate the individual function of production.

1.5.2 Decision-making and bargaining power analysis

Contrary to the New Home Economics Theory treating households as units of 'aggregated welfare' and 'rational choice' based on a unitary family utility function (Becker, 1981), the household nowadays is considered a place of divergent preferences among its members (McElroy and Horney, 1981; Chiappori, 1992; Manser and Brown 1980). This means that in the analysis, the utility functions of the spouses are taken into account and that the analysis is not restricted to the collective utility function of the household (Quisumbing *et al.,* 1999). The analysis of the bargaining power of husband and wife with respect to control over resources and decision-making concerning food security and household expenditures, consists of the following topics:

1. Assessing social and economic proxies of women's relative status (power) from reported questionnaires. We shall estimate the effects of women's power in decision-making on food and health provision, and joint income earned by husband and wife.
2. Estimating Engel functions for major categories of household consumption goods including expenditures on health, to test whether husband and wife pool their income to purchase goods. Pooling would imply equal propensities to consume from husband's and wife's income for each consumption category. No pooling would support the bargaining model (Phipps and Burton, 1998).
3. Estimating a bargaining model with no income pooling which considers husband and wife each maximizing their own utility (Himmelweit *et al.,* 2001, Lundberg and Pollak, 1993, Carter and Katz, 1997).

1.5.3 Analysis of health issues and household demand for health

The analysis of health issues will be accomplished within the framework of the micro-ecological approach to health (MEAH) (Niehof, 2004) and to the use of the Euroqol EQ-5D questionnaire (see e.g. McPake *et al.,* 2008). The MEAH approach emphasizes in particular two important considerations in analyzing health issues: one consideration is related to the socio-cultural factors embedded in the institutional environment (habits and attitudes in health-seeking behavior) and the second consideration is related to the way in which resources are allocated and used. We shall analyze the following topics: (1) the decision-making among household members regarding health care and expenses on health services across different types of households (landless, poor households), (2) households' decision-making regarding health care and ill-health costs, and (3) factors that have direct or indirect impact on health expenditures (Quisumbing *et al.*, 1999). The Euroqol EQ-5D is a standardized method for analyzing health outcome. It aims at depicting major constraints that may be induced by health problems and is applicable to a wide range of health conditions. The analysis of the different issues of this section will be made within the different chapters of the present thesis.

1.5.4 Welfare and well-being

We analyze the welfare and well-being of husband and wife by using the Leyden approach (Van Praag and Frijters, 1999; Kapteyn and Wansbeek, 1985) assuming that individuals are able to evaluate their income by using verbal qualifications. The estimated Welfare Function of Income (WFI) will be correlated with relevant variables, such as land type, personal income, etc. The well-being measures can be explained by taking into account a number of variables related to social and economic factors.

1.6 The study area

Senegal's population of 10.2 million has a density of 53.1/km^2. The estimated area is 196,192 km^2 lies between 12°8' and 16°14' North and between 11°21' and 17°32' West. Its west coast is the most western in Africa. The country comprises 11 regions and 34 departments. The research area has a dry Saharan tropical climate. Annual rainfall ranges from 400mm in the sylvo-pastoral North to 800 mm in the agro-pastoral South. The climate's principal characteristic is the shift between a long dry season of 8 months and an irregular rainy season of 3-4 months. During the dry season (October to June), temperatures can reach 47 °C. with a harmattan wind making human and animal respiration very difficult. At latitudes of 10-20° North, most of the rainfall occurs between June and the end of September. Ecologically, the area can be defined as a zone of semi-arid and shrubby land. Due to rainfall, natural flora also varies from desert to rain-savannah flora.

The research area can be distinguished by four types of soil: (1) skeletal soil not suitable for agriculture and with poor fodder value, (2) mineral hydromorphic soil, degraded soil used for pasture in the rainy season, (3) 'dior' soil mostly important, ferruginous and sandy, suitable for agriculture and land pasture with a variety of flora and (4) 'deck-dior' soil which is argillaceous and sandy, and used mainly for agriculture (Senegal, 2003).

The research work was done in two different areas: the sylvo-pastoral and the agro-pastoral zones. In the former, 50% of a farmer's income comes from livestock. In the latter, livestock is combined with cash crops and food crops and provides only 10 to 50% of a farmer's income. While the sylvo-pastoral area is characterized by a wealth of pasture land not suitable for growing crops, the agro-pastoral area is less suitable for grazing, and animal husbandry is mostly based on agricultural crop residues. The choice of these contrasting zones is relevant because it takes into account the variability in agro-ecological activities, support programs (e.g. NGOs, micro-credit organizations, extension programs) and facilities (hospitals, health centres, access to markets). Different ethnic groups in these areas are Wolof and Sereer (mostly farmers) and Fulani (mostly herders and some farmers). We shall discuss these issues more extensively in Chapter 2.

1.7 Outline of the thesis

The present thesis consists of seven chapters (Figure 1.1). Chapter 1 introduces the thesis, and Chapter 2 describes the field work and household characteristics with a differentiation across areas.

In Chapter 3, we model the allocation to agricultural activities of time and labour between household members, and the effects of socio-economic and cultural factors on total agricultural earnings. The model particularly specifies the effect of land property on total household earnings across different households within the same area and across areas. It analyzes the significance of gender participation in farming and non-farming activities.

Consequently, Chapter 4 analyzes farming in the sylvo-pastoral and agro-pastoral areas with respect to gender role and responsibilities concerning decision-making. In this chapter, differences in gendered decision making have been used to highlight actual behavior regarding practices and obligations in providing food security and caretaking in the household. Gendered valuation of ill health and decision-making concerning expenditures have been highlighted and captured. Moreover, Chapter 4 determines social and economic proxies of women's bargaining power and estimates the effects of these measures on food and health provision. Women's bargaining power is analyzed through their capability to participate in household decision-making (in terms of voice and management) and their share in household assets.

In Chapter 5 we assess whether husband and wife pool their income to purchase major household goods. Furthermore, a bargaining model with no income pooling has been used to estimate the behavioural attitudes of husband and wife towards consumption goods.

Chapter 6 analyzes the welfare and well-being of husband and wife by using the Leyden approach stipulating that individuals be able to evaluate their income using verbal qualifications. The estimated Welfare Function of Income was correlated with relevant variables, such as land type, personal income, etc. The well-being measures can be explained by taking into account an even larger number of variables like health, physical and social environment. Finally, Chapter 7 underlines the discussions of the major findings, conclusions and recommendations.

Chapter 2
Field work and household characteristics

Intrahousehold dynamics refers to the micro-economic analysis of the household and the relationship between husband and wife with regard to decision-making, bargaining power, allocation of resources and activities. However, important factors like norms, land use, access to services, credit and markets are important for a better understanding of intrahousehold dynamics. For this reason, the sampling was primarily based on differences related to households. Furthermore, access to revenues from agricultural or non-agricultural activities is a determinant of access to wealth and thus indirectly to good health. This chapter provides substantial information on (1) the sampling, (2) the data type and data collection, (3) the data management and analysis, (4) the Senegalese population, its agricultural production and constraints and finally on (5) the characteristics of health problems in the research area.

2.1 Sampling

Our areas of research comprise both the sylvo-pastoral and the agro-pastoral areas. Each area comprises regions with different departments. The departments comprise several communes divided into districts. The districts comprise rural communities divided into villages, hamlets and further households. In the sylvo-pastoral zone, the research study concerns the region of Louga and the district of Linguere (commune of Dahra-Linguere) comprising one district and four rural communities (see Table 2.1 and Figure 2.1).

In the agro-pastoral zone, the research is focused on the central agro-pastoral area of Senegal with two regions: Kaolack and Fatick. The departments concerned are Fatick, Gossas, Kaffrine and Kaolack. In October, 2005, the two areas were visited to gather preliminary information on the feasibility of the research work. It was particularly difficult to depict HIV/AIDS as an important health problem among the population encountered in the two areas, because they either did not know the disease or were not willing to talk about it. For this reason, in the analysis of gendered intrahousehold dynamics, one important recommendation for the analysis of health issues was to consider all health problems instead of focusing on HIV/AIDS.

The sample was based on the geographical and socio-economic diversity of the two areas. The agro-pastoral area concerns the central region of the 'Peanut basin' where 13 villages were sampled randomly. The sylvo-pastoral area concerns the District of Dahra with 12 villages. In total, 300 couples (husband and wife), either in two-person households or living with children and/or other adults, were randomly chosen from the village populations. The choice of the two areas was aimed to deal with three important considerations: (1) differences in access to agricultural activities,

Table 2.1. Research area: administrative units and villages concerned.

Area	Region	Department	Commune	District	Rural community	Villages and areas concerned
Sylvo-pastoral						
	Louga	Linguere	Dahra-Djollof	Sagata Djoloff	Boulal	Gouedia
						Boulylaobe
						Boulal
						Boily
						Belothiasky
					Deali	Wellouloumbel
					Sagata Djoloff	Thiangal
						Sagatta
						Thilla
					Thiamene Djoloff	Thiamene
						Deckwott
						Pampy
					Warkhokh	Warkhokh
Agro-pastoral						
	Fatick	Fatick	-	-	-	-
		Gossas	Gossas-Guinguineo	Mbadaxoune	Mdadaxoune	Mbadaxoune
				Ouadiour	Ourour-Ouadiour	Ouadiour
	Kaolack	Kaffrine	-	-	-	-
		Kaolack	Gandiaye	Koumbal	Latmingue	Thialane
					Thiare	Thiare
			Kaolack	Sibassor	Dya Ndiebel	Sibassor
						Ndiebel
						Kounkoung sereer
						Ndorong
			Khaone	Ndiedieng	Ndiedieng	Ndiakhatte
						Tallene
						Haffe
			Ndoffane			Ndoffane

(2) gender differences in access to resources and income, and (3) socio-economic and organizational aspects (access to services and extension programs, etc.) that can directly or indirectly affect household characteristics and resources. Although income may differ across areas, due to the factors mentioned above, analysis of variance showed that within the same area household income does not vary across villages. Because these three considerations were assumed to affect mainly women's role and productivity in the household, we expect that the two areas differ with respect to these

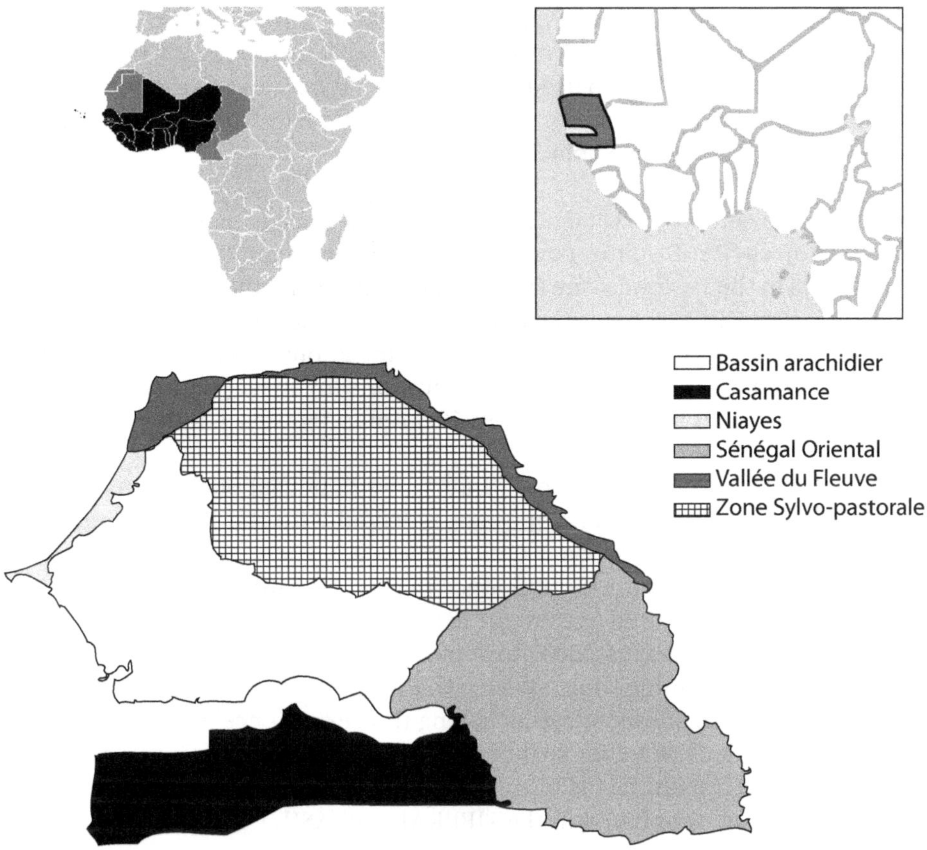

Figure 2.1. Maps: Senegal in West-Africa and location of the study area in Senegal. Legend: bassin arachidier = agro-pastoral (AP) area; zone Sylvo-pastorale = SP area.

various aspects due to differences in ethnicity, agro-ecological characteristics, land used and distance from cities. Besides, different farming systems are known to mean different livelihoods and different allocation of resources with respect to earnings and practices. From the choice of these two areas we expect that intrahousehold dynamics in terms of productivity, decision-making, women's empowerment and behavioural attitudes of husband and wife will differ across zones.

2.2 Data source and type

Data collection included two important aspects: collection of secondary data and the surveys. Information was gathered in different institutions through the study of reports and documents. For example, most information related to research into farming systems (production, marketing and trade) and agricultural constraints, was gathered through the Senegalese Agricultural Research Institute (ISRA). Relevant research work based on national reports in the agriculture sector benefited our preliminary investigation. I also used material from the Ministry of Agriculture, the Ministry of Livestock, the Ministry of Health, the Livestock Board Office and the health districts of Dahra and Kaolack.

Furthermore, relevant projects and organizations were visited, including the Livestock Assistance Project (PAPEL), the Poverty Reduction Strategic Project, the Union of milk Producers in the Pastoral Zone (UPPRAL) at Dahra and the Directory of Women for Livestock (DIRFEL) at Kaolack. Relevant information comprised statistics on agriculture and the livestock sector, farmers' organizations and constraints (climate, soil and rainfall), policies implemented, institutions involved (extension programs, credit and micro-credit institutions, intensification programs, gender policies).

Primary data was collected by personal interviews at household level using formal questionnaires. The household sample was selected with the help of some of the civil servants of the Livestock Board Office at Kaolack and of UPPRAL at Dahra, so as to ensure reliability of data and confidence in providing information. In fact, these civil servants later constituted a surveyor team involved in conducting the surveys. Data was gathered from August, 2007 to September, 2008. To accomplish the task in the survey period, six enumerators assisted in the application of the questionnaires: three in each zone. The enumerators have long-time experience in the application of questionnaires, work close to the farmers, and live in their area of work. Four of the surveyors are regular technicians of the Livestock Administration Service (DIREL) and the other two are employed in the UPPRAL and ASILA projects, respectively.

The questionnaires (Appendix 1) were divided into eight parts: (1) general characteristics of the household and its members, (2) gender activities and time allocation, (3) gender division of source of income, (4) Cost of production, (5) gender control over household income and allocation, (6) gender responsibilities in food provision, (7) women empowerment and bargaining power and (8) analysis of welfare and well-being. The general characteristics of the household were always elicited from its head. The other parts of the questionnaires concerning gender and time allocation to activities, bargaining power, health, welfare and well-being were administered to husbands and wives separately. 600 respondents (300 men and 300 women in the two areas) were visited to ensure the gender perspective approach. The health questionnaire (Appendix 2) comprises the Euroqol EQ-5D (McPake *et al.*, 2008) and the micro-ecological approach of health (coping strategies, decision-making regarding cost of health and care of ill members). Important topics in the questionnaires include:

- household composition;
- level of education and skills of husband and wife;
- social norms associated with gender activities;

- agricultural labour time, and activities performed by men, women and children;
- household labour time and activities performed by the household members both within and beyond their own household;
- time spent and activities performed by other people (paid agricultural labour and household work);
- individual ownership and allocation of land and livestock;
- individual income from market products (cash crop, food crop, milk and livestock products, cows);
- individual amount of borrowing;
- amount of individual food consumption and other costs including expenditure on health care;
- individual health status, individual evaluation of welfare and well-being.

Both at village and community levels, qualitative and quantitative data was gathered on issues like availability of services (extension programs and health-care facilities), availability of roads, utilities and market work, distance from markets. The qualitative data was gathered to support our quantitative analysis; several interviews with a group in each village (generally the village chief and some notable people) helped in understanding some issues such as major constraints in farming activities and trends in agricultural production. Most of this information is found in the present Chapter 2. With respect to the validity of the data gathered, the reliability of the enumerators is an important aspect. Since they all work in official services and benefit from long-time experience in extension programs and monitoring, farmers were confident in giving information. Moreover, the author was also involved in the application of the questionnaires in the two different areas.

2.3 Data analysis and data management

Descriptive techniques, such as analysis of distributions and differences between groups, and multivariate analytical techniques, such as multiple regression, were used to analyze the data (SPSS version 16 and STATA version 10). Theoretical models were set up to answer our research questions concerning expenditures from husbands 'and wives' incomes and whether income is pooled. Following this, a bargaining model was specified assuming different utility functions for husband and wife in a collective household model. In the health part of our study, a health demand function was estimated. The household health demand model intends to estimate an expenditure function related to health care in the household. Finally, we performed an econometric analysis using the Leyden approach concerning income evaluation to determine household welfare and well-being.

2.4 Population, agricultural production and constraints

2.4.1 Ethnic and demographic configuration of the households

In Senegal, the population is ethnically diversified, the majority comprising Wolof, Sereer, Fulani or Peulh (Halpular), Joola and Mandingue. In the areas studied, the main ethnic groups are Fulani mostly living in the Sylvo-pastoral area (SP) and Sereer and Wolof in the Agro-pastoral area (AP).

The Fulani or Peulh reside primarily in the North of the country, namely *Ferlo* and the south-eastern part. They represent 44% of the sample (73.0% of these Fulani live in the SP area and 16.6% in the AP area). They are primarily pastoralists (i.e. sheep or cattle farmers) moving from north to south, or semi-sedentary. They are normally herders and only few remain crop farmers (mostly millet crop). The basic Peulh producing system is livestock raising and herding. In fact, trading cattle and small ruminants is an important source of income contributing to the social status of the participating males. Production yields of livestock (milk and butter) are also very important as women's daily income. The livestock husbandry system is based on a cycle of movements depending on the season. These movements extend from the SP area to the AP area, where some Peulh have become sedentary over the years.

Wolof in the AP area are normally crop farmers. The food- and income-producing systems are generally based on cash crops, especially peanut. Growing millet and keeping livestock constitute secondary activities performed both by males and females. Generally, the Wolof language is used throughout the country. The majority of Wolof live in the cities Dakar and Thiés and in the major peanut areas of Diourbel and Sine Saloum. Wolof represent 37.3% of the sample. Of the Wolof in the sample 26.8% live in the SP area, and 47.7% in the AP area.

The Sereer represent approximately 18% of the sample. Living in the AP area namely the Sine Saloum, the central area of Senegal, Sereer practise a mixed farming system of crop farming and raising livestock. Food, especially millet, has an important social role (in redistribution between households) and a cultural (or symbolic) role in all life events like baptism, circumcision, wedding and death. In recent years, drought and climate uncertainty were economically disastrous to farming. For this reason, the Sereer tend to leave the rural areas for big cities like Dakar. Because of their proximity to such cities, economically speaking, Wolof and Sereer are more involved in non-agricultural activities than Peulh are. In fact the Peulh are a conservative ethnic group in terms of traditions and customs, and their seasonal movements mostly depend on water and feed availability for the herds.

2.4.2 Household size

In both areas taken together the average household size is 10.7 persons. As Table 2.2 shows, the average household in the AP area is larger than in the SP area (12 vs. 9 persons).This may be explained by the mobility of the Peulh which often implies segregation of household members in different areas due to scarcity of forage and water for animals. The average age of husbands (50) and first wives (37) shows a similar picture in the different areas. In the sample of the household members 52.3% are male and 47.6% female, which is in line with the sex distribution across zones: in the SP area the percentage of males in de population is 53.8%, whereas in the AP area it is 51.4. In the total sample, 51.2% of the household population is under 16, 23.6% are 17-25 and 18.7% are 26-45 years old. The remaining 6.5% are over 45.

Household size is also reported by ethnic group. In the Sereer group, households average 12 members, compared to 11 in the Wolof group, and 9.60 in the Peulh group (Table 2.3). However, the difference in household size between the different ethnic groups is not statistically significant.

Table 2.2. Household size and age of the spouses in the SP and AP areas.

	Minimum	Maximum	Mean	Std deviation
SP area				
Size of household	3.00	32.00	9.00	3.5
Age of male head of HH	17.00	76.00	49.87	11.52
Age of female spouses	17.00	73.00	38.50	11.54
AP area				
Size of household	4.00	30.00	12.41	4.45
Age of male head of HH	21.00	76.00	50.08	10.38
Age of female spouses	17.00	85.00	36.53	10.07

Table 2.3. Household size according to ethnic groups.

Ethnic group	Minimum	Maximum	Mean	Std deviation
Sereer	4.00	30.00	12.20	5.60
Wolofs	4.00	24.00	11.20	4.00
Peulh	3.00	32.00	9.65	3.80

2.4.3 Education

Five levels of education were distinguished for both males and females: (1) illiterate, (2) primary school not finished, (3) primary school finished, (4) secondary school not finished, (5) secondary school finished and (6) literate without training. The level of education was reported for 2,400 household members including children and relatives of the 300 households interviewed. Results show that in both areas, on average 51% of those interviewed are illiterate and 16% did not finish primary school. Among the 51%[1] illiterates, 75% are women. The analysis by zone shows a high level of illiterates in the SP area: 55% compared to 31% in the AP area (Table 2.4). Twenty-seven percent of the sample are 6 to 15 years old. Of those, almost 37% did not finish primary school, and 36% are illiterate. The important issue here is that the 37% who did not finished the primary school have abandoned the school (because of the burden of housework and the lack of monitoring by their parents)

The SP area has a lower percentage of educated people than the AP area. This aspect could be related to the ethnic groups living in the different areas. Table 2.5 shows that Wolof and Sereer are more educated and have a lower level of illiteracy.

This disparity between Wolof and Sereer on the one hand and Peulh on the other, may be explained by the fact that the latter are mostly nomads, moving around the country over the year. For this reason, they are often unwilling to accept any change in their direct environment (such as education, habits, etc.). In general, female spouses show a high level of illiteracy (81.64% in the SP area and 67.83% in the AP area). Non-educated wives represent 82% in the Peulh group, 70% in the Wolof group and 64% among the Sereer.

2.4.4 Structure of agricultural activities

Seventy percent of the Senegalese population are involved in agriculture and livestock farming. The latter constitutes an important component of agricultural production with a 70% share in Gross Domestic Product. In 2007 it involved 2.7 million head of cattle, 5 million small ruminants, 37,000 horses, 30,000 asses and 18,400 pigs. The share of livestock production in the research area accounts for 35% of national production.

The share of agriculture in national GDP decreased over the years, from 18.7% during the period 1960-1966 to 10% nowadays. Groundnut and cotton are the major cash crops in the rural areas. Food crops (millet, sorghum, beans and maize) occupy 60% of acreage per year and remain insufficient to cover annual demand (FAO, 1998; Senegal, 2004). The quantity of cereals consumed per person and per

[1] The average 51% concern all people in the households in our sample who are illiterate in the SP area and AP area (second row of Table 2.4) and those below 6 years (last column of Table 2.4)

Table 2.4. Education of household members (percentages of all people in the sample).

	Illiterate (1)	Primary school		Secondary school		Literate without schooling (6)	Illiterate below 6 years
		not finished (2)	finished (3)	not finished (4)	finished (5)		
SP area	55.13	12.87	6.52	2.48	3.10	14.86	5.02
AP area	31.25	19.58	4.69	9.58	3.84	20.36	10.69
Male heads in SP area	72.50	4.10	7.03	1.28	2.07	13.04	-
Male heads in AP area	34.80	6.46	2.38	3.46	7.81	45.09	-
Wives in SP area	81.64	3.96	2.62	1.40	1.58	8.80	-
Wives in AP area	67.83	6.72	1.81	3.46	1.38	18.80	-
6-15 years old	36.75	37.22	3.74	2.18	-	20.10	-

Table 2.5. Education according to ethnicity (percentage male heads and wives).

	Illiterate[1] (1)	Primary school		Secondary school		Literate without schooling (6)
		not finished (2)	finished (3)	not finished (4)	finished (5)	
Males, Wolof	52.43	5.58	2.82	1.45	3.72	34.00
Males, Sereer	55.75	9.45	3.84	6.60	6.36	19.57
Males, Peulh	77.00	2.60	2.65	1.98	1.80	12.40
Females, Wolof	69.72	4.63	2.84	0.95	0.92	20.94
Females, Sereer	64.27	5.72	3.83	7.50	3.83	13.85
Females, Peulh	82.00	3.75	1.54	1.50	2.22	9.03

[1] The illiteracy rate concerns only male heads and their wives interviewed in the sample (children excluded).

year is largely decreasing from 150 kg in 1970 to 70 kg in 2000. At the same time the consumption of rice is increasing, but the national production is insufficient to cover the demand (national production meets only 1/3 of the demand (Senegal/ FONGS, 2008). Aggregate agricultural production including cash and food crops has decreased by 40% over 30 years (Seck, 1997).

Land ownership

In Senegal, as in many West-African countries, land is an important issue. In the 'pre-colonial period,' the system was characterized by customary law by which land was the property of the family under the responsibility of the male head. Land was inalienable and daughters were excluded from succession. Later, colonial administrators tried to replace customary law land rights by property rights to the land, through personal registration of land; but without success, only 1% of land in Senegal was registered (Caverieniere *et al.*, 1988). After gaining independence in 1960, land was considered 'national domain' under the responsibility of the state. To farmers, land access was free and could be inherited. However, taking into account the gap between production and food security, recent reforms favour a redistribution of land for the benefit of better investments by rural communities and private persons. Recent political reforms in process tend toward increased recognition of farmers' property rights, and women's access to land should also be considered. In addition, taking into account the increasing conflict between herders and farmers, reforms tend to render more secure the interests of herders with respect to mobility and access to water.

In general, men and women work in the same plots of cash and food crops. However, to some women, their husbands allocated plots for food crops (mainly millet) or for kitchen gardening during the dry season. In our sample, 35% of wives cultivated millet plots during the year of the survey. Indeed, land availability is an important determinant of food and cash crop production but also in feeds and forage yields for livestock. In the sample, 14% of farmers have less than 1 hectare, 42% have between 2 and 8 hectares and the remaining 44% have more than 8 hectares available for growing crops. On average, 6.28 hectares are cultivated per household.

Agricultural activities

In the study area, household members are involved in many profitable activities in livestock and trade. Agricultural activities include food crops like millet, sorghum, maize and cash crops, particularly groundnut and water melon. The reports of farm activities show that females as well as males are involved in all agricultural activities, including sowing, transplanting, clearing fields, harvesting millet and groundnut and shelling the latter. Regarding livestock, detailed analysis in Table 2.6 shows that females are especially involved in milking, marketing milk, watering (bringing water

Table 2.6. Major livestock activities managed by household members (percentage of gender involvement).

Activities	SP area			AP area		
	Husband	Wife	Other	Husband	Wife	Other
Milking	11.20	74.22	14.58	13.33	14.15	72.52
Marketing milk	0.70	66.12	33.18	4.08	36.20	59.72
Watering ruminants	19.50	74.60	5.90	4.72	86.80	8.48
Fetching wood	35.64	63.00	1.36	40.10	54.32	5.58
Feeding (collecting and distributing forage and crop residues to animal)	54.50	42.70	2.80	39.34	3.33	57.33
Marketing ruminants	100.00	-	-	100.00	-	-
Guarding ruminants	53.00	3.40	43.60	4.00	0.70	95.3

to animals from outside the household) and fetching wood, while males are more involved in feeding, marketing and guarding ruminants. Nevertheless, females in the AP area are dynamic in trading or selling household goods at the market (12.2% compared to 0.7% in the SP area). These results suggest that major livestock activities are usually performed more by women than by men.

Regarding the participation of children in farming activities, results from the survey show that child labour is an important component of household resources. Even if the law provides for free education, many children do not attend school, and work on the farm instead. Indeed, the burden of household farming activities affects the rate of education in many areas. In fact children are involved in all tasks. They start work by the age of five and spend on average four to six hours per day; depending on the activity. Boys mostly work in the groundnut and cereal fields (53%), guard cattle and sheep (49%), collect water for ruminants (51.8%), fetch wood (41.0%), provide feeds (27.4%) and fetch water for household needs (24%) while young females are mostly involved in fetching water (59.0%), fetching wood (36.5%) and milking (16.0%). Estimates per zone show that in the AP area young children are more involved in farming than those in the SP area.

While livestock farming is the most important activity in the SP area, reported cultivation of food and cash crops is most important in the AP area (Table 2.7). In fact farmers in the SP area are mostly herders and less involved in growing crops. The better quality of land in the AP area is an important determinant of cultivation. Furthermore, the AP area is suitable for growing cash crops, especially groundnut. In most households, wives have their own cattle and sheep which are managed together with the household herd. In the whole sample, wives own on average five head of cattle and 11 head of goats and sheep. A description by area shows that these values are much higher in the SP area than in the AP area (Table 2.7).

Table 2.8 shows that the size of the household is positively correlated to land availability and land use (available land in peanut and millet). However, the correlation between growing crops and keeping livestock is significantly negative, meaning that high yields of peanut or millet are associated with smaller numbers of sheep and goats. This result can be illustrated in the AP area where the increase of crop yields takes priority over expanding the number of sheep and goats, thus reducing the acreage needed for the latter. This reflects a specialization of farmers in the different areas.

2.4.5 Constraints and strategies

Nowadays, the major causes of the deficiency of agricultural production can be related to the climate (rainfall decrease of 35-40% in the North and 20-25% in the South during the period 1970-2000) and to decreasing soil fertility over the years, due to overexploitation of land and to inappropriate agricultural policies (insufficient

Table 2.7. Descriptive statistics of crops and livestock.

Input factors	SP area		AP area		Total sample	
	Mean	Std	Mean	Std	Mean	Std
Land available	7.40	0.50	10.0	0.61	8.70	0.40
Land peanuts	0.64	0.12	3.43	0.26	2.00	0.16
Land millet	1.10	0.14	3.64	0.21	2.30	0.15
Land beans	0.46	0.06	0.36	0.06	0.41	0.04
Land maize	0.03	0.01	0.85	0.10	0.44	0.05
Land water melon	0.20	0.05	0.99	0.21	0.60	0.11
Number of cattle/HH	34.02	35.00	15.86	17.48	24.90	29.05
Number of sheep/HH	84.95	27.22	9.33	12.82	46.32	99.37
Number of goats/HH	28.03	129.31	7.95	10.56	18.65	22.61
Number of horses and donkeys	7.74	13.95	2.73	1.84	5.21	10.20
Wife ownership cattle	8.87	10.92	0.92	2.72	5.00	8.89
Wife ownership sheep/ goats	20.09	29.57	1.93	15.39	11.00	23.02

Land size is in hectares (ha); HH= household.

Table 2.8. Correlations between agricultural variables and livestock resources.

	Total land size cultivated	Land groundnut	Land millet	Number of sheep	Number of cattle
Size of household	0.327**	0.365**	0.384**	-0.112	-0.022
Total land size cultivated		0.673**	0.706**	-0.058	-0.022
Land groundnut			0.688**	-0.188**	-0.101
Land millet				-0.162**	-0.078
Number of sheep					0.473**
Number of cattle					

** $P<0.01$, * $P<0.05$.

extension programs, abolition of subsidies, high price of inputs, deficiency in trade policies and unavailability of good seeds, among others). Moreover, the liberalization of markets has been accompanied by a disengagement of the state (from providing subsidies and credit). As a consequence, with an increase of devastated vegetation cover in many regions and a decrease in production, farmers and herders have developed survival strategies. Among these strategies are several income-generating mechanisms (Macintosh, 1989; Fall *et al.*, 2004; Ware, 1979):

- Traditionally maintained as a prestigious activity, through 'beef and sheep fattening' livestock is now a commercial investment and a strategy to generate quick income.
- The increasing number of young males in the household in search of full-time and non-farm employment in the towns and cities, has consequently increased the abandonment of farming and decreased yields of cash crops and food crops (Dieng, 1998).
- Men's and women's seasonal exodus to the cities (Dakar, Thiés, Diourbel and Touba), in search of a secure source of income resulting in later cash income transfers to the household.
- Widespread selling and pawning of farm tools and equipment has led to their quasi-absence in households.
- Growing water melon as an alternative to deficient peanut crop production.
- Increasing non-farm activities like crafts industry and tree exploitation.
- Increasing commercialization of milk by women and, recently, by men (intensification and cross-breeding).
- The rapid growth of the informal sector (trade, services).
- The proliferation of micro-projects and women's networks addressing common issues focused on income generation.

The seasonal exodus to the cities has stimulated other independent activities like entrepreneurship, running a shop or employment in administration or craft industry (Table 2.9). One other important consideration for both men and women in the AP area is the possibility of small-scale paid work in cities. For example, 10.8% of females in this area work especially in the dry season as home maids in peri-urban towns or in cities. In fact, in recent years, food crop yields generally attributed to women have been reduced or abandoned because of drought and lack of inputs (fertilizers, seeds).

2.4.6 Agricultural extension and farmers' access to other services

Since the event of liberalization and the withdrawal of the state from supporting the agricultural sector, the outreach of the rural extension services has been declining with respect to those services and programs involved in the development of agricultural activities. The substantial decline of extension services has been largely responsible for the decline of livestock and crop productivity. However, there are still some government agricultural services available from the Ministry of Livestock (district livestock department) and the Ministry of Agriculture (district agricultural department). Also, some projects and programs funded by the World Bank, by the Development Bank for Africa (BAD) and by some NGOs or private suppliers of inputs and services aim at supporting the development of the agricultural sector but are only available in some districts, thus limiting the access of most farmers to agricultural services. Mostly, these services target their support at promoting intensification through better inputs (seeds and crossbreeding) or feeding strategies for livestock and access to credit and micro-credit. For both livestock and crops the intensification programs aim at increasing production of milk and meat, a better procurement of crop production and a diversification of activities (kitchen gardening, processing and trading). Among the institutions intervening in our area of research, one can mention the Livestock Assistance Project (PAPEL) managed through the Livestock Board office (DIREL), the National Agency for Rural and Agricultural Advice (ANCAR) and other institutions of micro-finance. The DIREL and PAPEL

Table 2.9. Farm and off-farm activities (percentage of gender involved).

Activities	SP area		AP area	
	Husband	Wife	Husband	Wife
Farm activities	78.22	82.33	72.84	48.10
Trade	4.42	0.73	8.65	12.24
Employer (teacher, home maid)	2.00	3.50	5.30	12.80
Other (entrepreneur, shopkeeper etc.)	15.36	13.44	13.21	26.86

programs have several objectives for livestock production: (1) the preservation of the pastoral livestock system in traditional dry areas managed by the Peulh is the most important one in Senegal, (2) the modernization of livestock through various measures (feeding strategies, access to credit, water supply, supplementation), (3) better access to facilities like infrastructure for the distribution of products, e.g. transportation, (4) safeguarding the marketing system in order to improve livestock income, (5) assistance of farmers' organizations (in terms of capability reinforcement, education, micro-credit access) and management and (6) assistance in the form of advice to farmers.

ANCAR has also implemented some pilot projects in several districts promoting a better access of farmers to productive technologies in horticulture, agriculture, fattening, diversification, and conservation of the environment. For the special case of local milk production, modernization is based on the assistance of organizations involved in milk production and collection processes in rural areas and on the development of intensive peri-urban farms (production based on cows with high genetic potential). The crossbreeding based on artificial insemination is subsidized by the government at more than 80% of the cost. Among other national or private institutions providing extension community services in our area of research, we can also list the Mutual Agency of Senegal (Caisse Mutuelle du Senegal (CMS)), the Senegalese Agricultural Bank of Credit (CNCAS), the Hunger Project, Caritas, the Enterprise for Textile Fiber Development (Sodefitex), the Village Organization Program (POGV), the PROMER and also many mutual agencies involved in micro-credit (Djomec, CAPEC, PMIA and Pamecas, etc).

In recent years, farmers' associations played an important role for government services and private institutions involved in extensions and micro-credit. We mention UPRAL (Farmers' Union for Milk Production) and the DIRFEL (Directory of Women in Livestock), MDE (Herders' House) and many other small groups. The small groups are present at the village level while the large associations (MDE, UPRAL, DIRFEL) are active at the regional level. The group of farmers at the village level is generally based on financial contributions of farmers in the form of savings. The women's mutual agencies are particularly dynamic in providing micro-credit. Table 2.10 shows that almost all villages visited have access to some support such as credit or extension services. However, the number of services involved varies between villages. In general, remote villages have less access to community and extension services.

A comparison between areas shows that the AP area benefits more than the SP area from extension services and credit. In fact, the AP area is the focus of most programs because of the important role of cash crops in the national economy (groundnut, cotton and water melon). The availability of roads and other infrastructure and the relatively shorter distance from Dakar (the central city) may also account for this.

Table 2.10. Percentage of villages in which different community services are provided.

Type of services	SP area	AP area	All samples
Government agricultural extension services	10	40	35
National projects	25	45	45
Agricultural National Bank of Credit	15	40	20
Micro-credit institutions	35	40	65
Agricultural research institution	10	2	10
Private providers of services	20	30	25

Regarding access to credit, in the AP area 40% of villages have access to the National Bank of Credit, compared to 15% in the SP area. In the AP area 23.20% of women have access to micro-credit, compared to 20% in the SP area. The Agricultural National Bank of Credit (CNCAS) focuses its attention predominantly on the head of the household, in most cases a male.

2.5 Health situation in the research area

This section introduces the health issue in our study as a major shock that may induce vulnerability to people, especially poorer people. The first part will describe the general health situation in Senegal, and the second part will provide additional information based on our surveys using the Euroqol EQ-5D questionnaire.

2.5.1 Overview of the health situation in Senegal

Despite the remarkable progress assessed through health indicators like mortality and morbidity rate in some countries like Senegal, health problems remain an important cause of low income and low productivity (Oxaal and Cook, 1998; WHO, 1995). It has been recognized that health problems have important linkages with gender (allocation of time and labour) and poverty. The gender issue is critical because of the impact of malnutrition and decision-making regarding the allocation of time and labour, and the access to health care and health facilities and socio-economic constraints. In the past decade, policies have focused on improving health status because of its important role in economic productivity. These policies have especially emphasized women and children because of their special needs. In poor developing countries, the challenge has been worse because of substantial inequalities in their access to basic services (water, health facilities, low quality of assistance and equipment). Health-sector reforms have put more emphasis on the gender issue through specific programs, poverty and inequity. For example, the view on health as a basic right has led policies and governments to make great efforts in the provision of basic services for the poor through appropriate and low-cost services.

The Senegalese health condition has really improved in recent years. However, important constraints related to the access to health centres and the high cost of medical health care still remain. The Assessment Poverty report (Senegal, 2007) shows that overall 60% of households can reach a health service within a 30 minutes' walk; but about one in four households are one hour's walk or more from a health service. These difficulties related to the proximity of services are more pronounced in rural areas than in cities. Among other difficulties reported at the national level, 9.4% of people who use health centres consider the cost to be very high, 4.3% judge the health structures to be inefficient and 3.3% complain about the long waiting times. The Senegalese Demographic and Health Survey (MEF, ANDS, 2005) published detailed information related to the anthropometric and medical situation in rural and urban areas. Surveys realized all over the country underline the higher vulnerability of children and women. Around 21% of children suffer from malnutrition and 84% of under-fives have anaemia, while the percentage of women with anaemia is around 61. The prevention of child diseases has improved, because 70% of children have a vaccination booklet and among them 59% have completed the vaccination scheme (within the 12-33 months age bracket). Between 1992 and 2005 vaccination coverage increased from 49 to 59%. However, the HIV/AIDS epidemic has increased the vulnerability of people to existent diseases. Considerable effort has been made by the government and other institutions with regard to increasing people's awareness. The EDS-IV' surveys (2005) indicate that on average 97% of people know of the disease. The screening test for AIDS reveals 0.7% prevalence at the national level (on average 0.9% for women and 0.4% for males). The prevalence results depict a higher vulnerability of women compared to men: for every 100 males infected, there are 225 infected women (ratio of 1:2.25). The prevalence of and people's knowledge about AIDS (mode of prevention, type of disease) depends upon the area of residence and the level of education (MEF, ANSD, 2005). Since 2004, the government has implemented the National Plan for Social and Sanitary Development (PNDS) that prioritizes several improvements within the health sector.

2.5.2 Health status in rural Senegalese households: the Euroqol EQ-5D approach

Results from our sample provided relevant information with respect to husbands' and wives' health status. Affections reported in case of acute disease are mainly related to malaria (65.1% of cases), flu (18.7%), diarrhoeic or stomach symptoms (6.4%), and other problems (reproductive problems of women or injuries). According to interviews with farmers, the consequences of these acute diseases are anaemia and mortality especially if not treated. Chronic health problems often are related to diabetes and other unknown diseases or malnutrition (2% of cases). Symptoms clearly related to HIV/AIDS, even though existent, were not reported in our surveys. Non-detection of HIV/AIDS problems may be due either to silence of the members interviewed or by their lack of knowledge regarding the disease. The Euroqol EQ-5D questionnaire, related to the direct impact of illness on individuals' performance, includes five indications of husbands' and wives' health status according to McPake *et al.* (2008). The indications are related to factors that might affect individuals and impact on their health status in case of illness. The factors are related to (1) the mobility of people, (2) the capability to carry out usual activities, (3) the presence of pain, (4) the presence of anxiety or depression, and (5) the availability of care and treatment in case of illness. Data from the survey shows that respondents do not report frequent illness problems that might affect their mobility (Table 2.11). Eighty-one percent of women and 71.40% of men do not have such problems. However, 22% of respondents have moderate problems of mobility. During the period of the survey 63.39% of wives and 60% of husbands did not report health problems that affected their usual activities. However, on average 8% of the sample reported incapacity to perform their usual activities (10% was reported by husbands and 6% by wives separately. With respect to pain and anxiety in case of illness, 43% of husbands and and almost the same for wives feel moderate pain. These estimates show that individuals affected by illness report problems related to their capability or performance and to their quality of life (anxiety, pain). Husbands and wives interviewed separately show almost the same estimates with respect to these problems. When we look into possible access to treatment, almost 80% of respondents report no access compared to 15% reporting little access to health treatment. These observations show that illness problems may be correlated to productivity and decision-making in the household. For this

Table 2.11. Some indicators of the health problems of household members (in percentages).

Euroqol EQ-5D Indicators	Husband	Wife
Mobility constraint		
No problem	71.48	80.81
Some problems	25.84	18.18
Incapacity	2.68	1.01
Performance of usual activities		
No problem	59.60	63.39
Some problems	29.63	30.51
Incapacity	10.17	6.10
Pain and comfort		
No pain	47.47	48.81
Moderate pain	43.10	42.71
Extreme pain	9.43	8.47
Anxiety and depression		
Neither anxiety nor depression	72.48	76.43
Moderate anxiety or depression	25.50	22.56
Extreme anxiety or depression	2.01	1.00

reason, the health status of husbands and wives is used in the remaining chapters to capture the effects of health problems on rural farmers with respect to welfare and well-being.

2.6 Conclusion

A description of the households in the survey shows a diversity of ethnicity and farming systems in Senegal. While the SP area is mostly populated by Peulh (73% of the sample) and Wolof (27%), the AP area is mainly inhabited by Wolof and Sereer. The average household size in the AP area is 12 persons, in the SP area it is 9. The latter area also represents a low level of educated people: 55% illiterates compared to 31% in the AP area. In addition, females in the two areas show a high level of illiteracy: 81.64% in the SP area and 67.83% in the AP area. Land as an important and basic asset for the household remains crucial to agricultural production and women have no property rights to it. Land availability varies across households and 14% of the farmers have less than one hectare. In fact, land availability is an important factor of production, a topic that will be discussed in Chapter 3. Household members mostly work in mixed farming systems where agriculture and livestock are complementary activities. The number of cattle and sheep co-determines income generation through milk and meat production. For example, marketing milk by women favours daily income in the household, and earnings from sheep constitutes an important part of permanent household income for husbands as well as for wives, especially during some important events (Aid El Kabir and Aid El Fitr). Even women do not intervene in marketing livestock at the market, they nevertheless keep the earnings from their livestock assets. The agricultural sector produces cash crops like groundnut and cotton, and food crops like millet and beans, mostly in the AP area. Nowadays, water melon also constitutes an important source of earnings because of its adaptation to crop cultivation planning. Further details on these issues will be discussed in Chapter 3. Men, women and children are all involved in the farming system, in all tasks but at different levels of performance. Women and children actively contribute to agriculture and livestock activities.

Various forms of extension services are involved in the agricultural sector aiming at its intensification. However, their accessibility is limited for most farmers, especially those living in remote areas. For this reason, nowadays there is a trend in the proliferation of farmers' associations and groups in all areas.

Results show that a priori, females and males participate differently in the labour market and other activities. Females are involved in many activities as farmers or traders and also as employers in small-scale activities in Dakar or other cities. These activities have several consequences for their labour time allocation, for the productivity of farming and for decision-making regarding expenditures.

The description of health problems shows that illness may induce negative effects at the level of the household both for husband and wife. While substantial effort has been made in care and access to health facilities in rural areas, access to health centres and treatment is constrained, especially in remote areas. Moreover, people face difficulties in accessing resources (capital) to finance direct costs related to health treatment and health care. Results from our survey showed that health problems result in direct negative effects on individuals with potential impact on productivity and well-being. In-depth knowledge will be provided with respect to decision-making concerning labour and time allocation and health expenditures in the household. These different aspects will be analyzed in the remaining chapters.

Chapter 3

Gender, resource allocation and productivity

3.1 Introduction and theory

3.1.1 Introduction

Agriculture is the primary source of income in Africa. In Senegal, it involves 70% of rural people. However, Senegalese agriculture is confronted with low production leading to food insecurity and increasing poverty. The social, cultural and economic factors that may affect household productivity mould the way in which men and women benefit from agricultural production. Research in Africa has been focused mainly on the impact of technologies for intensifying production, or on land tenure's impact on household productivity, and has rarely identified intrahousehold factors and gender roles that may affect household production. Different sectoral growth rates in agricultural production (cash crops, food crops and livestock) induce different demands for women's and men's time allocation to agricultural activities (World Bank, 2005; Lilja *et al.*, 1998). For this reason, it is relevant to analyze the gender roles in production and their determinants. Understanding human and physical resources as well as socioeconomic and community characteristics that may affect agricultural production at the micro level, may help in identifying constraints and has a positive impact on designing adequate programs and projects. The purpose of this chapter is to contribute to this issue and to provide empirical findings regarding the importance of gender allocation of labour and time in agriculture, and the identification of individual, socio-economic and cultural factors that may affect individual and household earnings.

3.1.2 Theory and relevance of the main findings

The debate about the efficiency of agricultural activities in the household concerns especially the household resource allocation, in particular labour time and human capital allocation (Kandler, 1988, Lado, 1992; O'Neill and O'Neill, 2005). The traditional gender division of labour within the New Household Economics framework was viewed as a rational economic response to market valuations of people's time, reflecting their productivity in the market. Becker stated in 1965 that '*Members who are relatively more efficient at market activities would use less of their time at household activities than would other members*' (p. 512). Assuming that households make choices about the allocation of time to maximize their utility, Becker's explanation of gender division of labour was based on the specialization of men in market production and women in household activities because the men earn higher wages.

Furthermore, New Home Economics points out that labour specialization has consequences for the earnings of spouses, suggesting that the one who performs less market work, earns less. In the context of developing countries, studies have investigated the allocation of labour and time especially for females and children, to explain the effect of human capital on earnings. Some evidence (Brown and Haddad, 1995; Gallasso, 1999) underlined the special role of women and children in work activities at the farm level. The contribution of males and females to farm activities appeared to be guided by a specialization of the different agents (Fafchamps and Quisumbing, 1998; World Bank, 1999) and by seasonality (Alderman and Chishti, 1991). For example, though women are working hard in farming activities, their earnings are minor and their role in the household has been often associated with household chores. In contrast, males were seen as the main breadwinners. Even though spending time and labour have been associated with higher income, it has been argued that men are more productive in farming activities and earn more than women in market work. Moreover, attempts have been made to provide more insight into the question whether women's time allocation is subject to economic constraints as suggested by the economic theory of the household (Becker, 1965; England and Farkas, 1986; England and Kilbourne, 1990; Haddad *et al.*, 1997). Nowadays there is a growing literature on time use, particularly in developing countries, because of increasing poverty and declining earnings. Some studies in developing countries have provided empirical evidence of gender specialization in farming activities, showing females to be more specialized in livestock production (Jacoby, 1991). Also, results from Ivory Coast revealed that female labour had a large share in profits from growing crops (Quisumbing, 2007; Jacoby, 1995). Since time use is a variable factor in the household, it might explain the variability of resources and earnings among household members and across different geographical areas. In fact time allocation as a farm resource can vary among people and across other household resources (assets, fixed inputs). However, while a number of studies (Becker, 1965; Ciccone, 1994; Stokey, 1991) examine the effects of labour allocation on agricultural output, the allocation of time within different categories of people (for example, landowners and the landless, men and women, girls and boys) has rarely been analyzed. Some studies have focused on the links between labour allocation, productivity and responsibility (Kimhi, 2000; Fafchamps and Quisumbing, 1998) or between labour allocation, productivity, social networks and norms (Kevane and Wydick, 1999; Xinyu *et al.*, 2007). These studies were mainly focused on aggregate productivity at household level, thus ignoring gender differences and differences across activities.

A better understanding of time allocation of household members by gender, would probably help in implementing suitable projects or extension programs with respect to production and total earnings from agricultural activities. Also, we expect that livestock and crop-growing activities are interdependent on a gender specialization within activities regarding time allocation and labour. In addition to our expectation that women's and children's labour has a positive association with farm production, our first hypothesis is that female time allocation to farming is negatively associated with time allocated to housework and caring.

In the context of developing countries, the imperfection of markets has a great influence on the productivity of different household members. We expect that total earnings which is an important indicator of production is affected by socio-economic, individual and structural factors. For example, the access to land and assets (livestock, capital endowment) is an important factor in securing revenues from farming activities. In most developing countries, access to land is a critical issue in income-generating activities and capital accumulation especially for women (Gordon, 1996; Davis, 2003; Bushal, 2006). In addition, lack of information on the market and constraints related to physical inputs (land, livestock) or socio-economic factors (lack of capital endowment, access to credit or micro-credit) could disfavour farmers' earnings (Mayoux, 2001; Ao Onoya, 2008). Particularly, the lack of information about new agricultural technologies is directly linked to agricultural performance of husband and wife and to absence of knowledge about credit or micro-credit. With respect to individual performance, the lack of knowledge may be related to the low level of education that can limit investments, thus reducing earning opportunities. In fact, education as a human-capital factor with a positive effect on productivity, has attracted a lot of attention in empirical work (Lucas, 1993; Stokey, 1991; De Brauw *et al.*, 2008). However, in the context of developing countries, results vary according to communities and regions (Verwimp, 1999; Winters *et al.*, 2008). In addition to socio-economic factors, we intend to investigate whether education is associated with male and female earnings. Also, the ongoing literature, considering whether women's time allocation is also affected by local norms or patriarchy (Gordon *et al.*, 1996, Kevane, 1998; Moghadam, 2004), underlines that cultural factors may handicap female labour supply in income-generating activities outside the household. Observations and insights advanced by researchers show a gender-based division of labour in developing countries where women are forced to allocate their time to cash crop production work (Elson, 1991; Darity, 1995) and receive less return. In our sample, we must distinguish different ethnic groups involved in farming activities. The SP area is mostly occupied by the Peulh, a pastoralist and conservative group of people who are less educated than Sereer and Wolof, who live mostly in the AP area. This situation can lead to women's contribution in farming being even further influenced by cultural norms. Hence, we test the hypothesis that household production is associated with individual, household and community characteristics. In doing so, one relevant issue will be the understanding of major factors that can be correlated with the productivity of agricultural activities.

Since agricultural activities are all linked at the household level (growing cash or food crops, keeping livestock), this study will investigate not only how the intrahousehold allocation of human capital and time affects production estimated as total earnings, but also the way in which households with different types of resources and activities (livestock, crop-growing and non-agricultural activities) are associated with individual, household and socio-economic characteristics. The first part of this chapter comprises a descriptive analysis of the relative importance of farm and off-farm activities and earnings differentiated by gender. This part (Section 3.3.1) also discusses time spent on the different activities. The second part of this chapter (Sections 3.3.2, 3.3.3 and 3.3.4) models the production of the household and deals with individual income and individual time allocation to farming and its determinants.

3.2 Method

3.2.1 Definition of concepts: production, productivity and efficiency

In our research work, the rural households derive their livelihood from several income-generating activities. Family members perform agricultural activities that account for much of production. Some family members work outside the household for a wage. Production then depends on the interdependence of family members' activities in the utility and household production function. So, total production depends on family labor supply to the household activities. In the following sections, we refer to total production as estimated total earnings from crop and livestock activities (see Section 3.2.2).

On the other hand, productivity is defined as the ratio of a volume measure of output to a volume measure of input use. The concept of productivity then implies in the context of our study, the consideration of total output estimated through the total earnings related to physical inputs (land), labor and time allocation. There are many different productivity measures; the choice depends on the purpose of productivity measurement and in the availability of data (OECD, 2001). The notion of productivity also leads to an important concept called efficiency.

In the theory, the level of technical efficiency of a particular enterprise is characterised according to Greene (1993) by the relationship between observed production and some ideal or potential production. Moreover, efficiency means that a production process has achieved the maximum output with a given fixed input (Diewert and Lawrence, 1999). However, economic efficiency refers to either technical efficiency or allocative efficiency. The latter refers to an optimal allocation of resources implying profit-maximisation.

Rather than studying efficiency in this chapter, we are interested in measuring productivity as the relationship between gross output (total amount of household production expressed as earnings without subtracting the cost of inputs) and inputs used. For example, to the extent that labor supply raises productivity, we expect significant positive correlation between this factor and total household output (expressed in earnings), after controlling for other inputs and semi-fixed factors. In the estimation of the total earnings derived from total production (total production times price for all products), we have included the own consumption of the household, particularly regarding food crop and milk production which constitute an important part of household consumption. In fact, the lack of reliable data on household self-consumption of own production has not allowed us to describe and desegregate the total household earnings into the total amount of household products sold in the market and the total self-consumption of products. The total household earnings derived from livestock and crop separately (see Section 3.2.2) takes into account the aggregate value of household production in franc CFA.

3.2.2 Method

The sample includes 300 rural households in which husbands and wives are married couples and earn income primarily through agricultural activities, and secondly, through non-agricultural activities or cash transfers. Since we assume that production is strongly linked to earnings, we report all activities both within and outside the household, and as our measure of production we assess income generation per agricultural activity during the year of survey. Couples have been interviewed separately using questionnaires, and income has been reported for each farm activity including livestock, cash crops, food crops, vegetables, transfers and off-farm activities. Since time and labour are production factors, analyses will concern resources of all household members (wife, husband, other active household members including children). The analysis aims to depict the correlation between gender participation in market and home production and the labour time and human capital associated with different activities (among farmers with different land-owning status). We assume that production from farming activities (growing crops, milk production, and livestock) and housework depends on a number of variables including labour, human capital, skills, status, variable and fixed inputs, and other characteristics of the household. The following general equation will be specified:

$$Y_{xi} = G_{xi}\,(L_{xi}, C_{xi}, V_{xi}, F_{xi}, H_{xi}, Z) \qquad (3.1)$$

where Y_{xi} denotes earnings from activity x (crops, milk, livestock) of individual i (husband or wife) without subtracting the costs related to the activity. G_{xi} is a production function associated with activity x of individual I (husband and wife), L_{xi}, an individual's labour time devoted to x, C_{xi}, time devoted to homework for the wife in the household, V_{xi}, variable input, F_{xi} fixed input (land) associated with activity x of individual i, H_{xi} denotes human capital of the individual (skills, education) and Z is a vector of household characteristics. The distinction of variables by gender allows us to specify several models:

- Husband and wife equations to determine the effects of important factors on livestock and crop output and their variations across zones.

- Husband and wife equations to determine the effects of individual, household and community characteristics on husband's and wife's income, from crops and livestock, respectively.
- Husband and wife equations to model time allocation in farm work.

Cobb-Douglas functions have been specified for these models, based on several factors that potentially may be correlated to the household's output[2] from farming like household labour, land use, time allocation, home work involvement, human capital variables and other inputs (including off-farm earnings and cash transfers). Since we express the production of agricultural activities in total earnings as dependent variable in most cases, we only focused on the correlation between some important factors and this dependent variable and did not analyse causality or reverse causality. We use our common sense and the economic theory to assume the correlation between variables. We assume that labor and time allocation to agricultural work and housework as well as other factors (livestock assets, land, access to credit and so on) have an effect on total earnings and that the reverse causual effect is hard to tell and is not analyzed in our discussions. For example, it would be hard to explain that more earnings induce more time devoted to crop or livestock activities because of the inelasticity of the availability of some important input such as land ownership (that cannot be acquired in the market). May be more earnings can lead to a substitution of the labour supply through, for example, the hire of labour and the increase of the leisure time or the time allocated to other non agricultural activities (off-farm jobs) of husband or other household members. The same reasoning can be applied also to housework, access to credits and tranfers from migrants: do more earnings lead to more housework and to more access to credits? Since we deal with cross-sectional data and do not have instrumental variables such discussion regarding causality may not be reliably be incorporated in our findings. However, besides the estimation of the statistical significance of the different variables on household earnings in farming we also have estimated the economic significance of some effects by computing the beta coefficients. Because of all our variables are not all of the same scale, we can then rank the regressors by the magnetitude of their beta coefficients. The absolute magnitude of the beta coefficient of the independent variables indicates the strength of the effect of that variable and hence its economic meaning.

For livestock in the total sample, essential inputs concern labour (L_b), size of livestock herds (L_v) (determined by the number and type of animals, time allocation (T) in livestock and in homework, the number of draught animals (D_a) which contribute to transportation of feeds and water, and land (L_d) which supplies fodder, forage and

[2] In our analysis we consider that there is a relationship between dependent and independent variables, although we neither assume nor test for causality of the relationship. Only for the sake of convenience we speak of independent variables 'affecting'or 'having an impact'on the dependent variable, where technically speaking we mean no more than 'association' or 'correlation'.

crop residues for animals. The human-capital variables are related to education, age and ethnicity of household heads; other inputs are summarized into *Z;* ethnicity and education are categorized as dummy variables. Z comprises among other variables, health status and distance from the market, that may be related to household and individual earnings. Total earnings from livestock (Y_l) can also be affected by a normal error term *u*, added to the equation. The livestock production function can be specified for each household as[3]:

$$Y_l = \alpha_{0l}\, L_b^{\alpha_{1l}}\ (1+L_v)^{\alpha_{2l}}\ (1+L_d)^{\alpha_{3l}}\ (1+D_a)^{\alpha_{4l}}\ (1+T)^{\alpha_{5l}}\ e^{\alpha_{6l} z}\ e^{u_1} \qquad (3.2)$$

Taking logarithms of both sides, Equation 1 can be written as:

$$\ln Y_l = \ln \alpha_{0l} + \alpha_{1l}\ln (L_b) + \alpha_{2l}\ln(1+L_v) + \alpha_{3l}\ln(1+L_d) + \alpha_{4l}\ln (1+D_a) + \alpha_{5l}\ln(1+T) + \alpha_{6l}Z + u_l \qquad (3.3)$$

For crops, the same reasoning can be applied for the estimation of the model. However, both land and number of draught animal employed have a direct effect on crops. The size effect of livestock is also taken into account since in the context of developing countries livestock activity is considered complementary to agricultural activity and it is often used to provide an organic way of saving assets. The equations for crop earnings can be written as[4]:

$$Y_c = \alpha_{0c}\, L^{\alpha_{1c}}\ (0.1+L_d)^{\alpha_{2c}}\ (1+D_a)^{\alpha_{3c}}\ (1+T)^{\alpha_{4c}}\ (1+L_v)^{\alpha_{5c}}\ e^{\alpha_{6c} z}\ e^{u_2} \qquad (3.4)$$

$$\ln Y_c = \ln\alpha_{0c} + \alpha_{1c}\ln L_l + \alpha_{2c}\ln(0.1+L_d) + \alpha_{3c}\ln(1+D_a) + \alpha_{4c}\ln (1+T) + \alpha_{5c}\ln(1+L_v) + \alpha_{6c}Z + u_2 \qquad (3.5)$$

Since zero income was noticed for some 9.7% of the livestock sample and 49% for crops, and to avoid selection bias, we estimate the equations using a more general model which allows for selection of having output, being determined by variables different from those that determine the values taken by the positive observations. This approach uses the Heckman two-step selection model which estimates total outputs in all equations as censored variables (Heckman, 1976, 1979). The Heckman selection model uses two equations. The first general equation captures Equations 3.2 and 3.5, and is called the outcome equation:

$$Y_l = Y_1(\alpha' X + U_1) \qquad (3.6)$$

Where $\alpha' X = \alpha_1 X_1 + \alpha_2 X_2 + \ldots \alpha_n X_n$, and where Y_1 is observed and U_1 represents the error term. The second general equation is the selection equation:

$$Y_2 = g' Z + U_2 \qquad (3.7)$$

[3] We add 1 to the explanatory variables in order to avoid computational problems when taking the logarithm.

[4] We add 0.1 to the explanatory variable for land since most households cultivate 1 hectare.

Where $g'Z = g_1 Z_1 + g_2 Z_2 + ...g_n Z_n$

$Y_2 = 0$ if $Y_2^* \leq 0$

$Y_2 = 1$ if $Y_2^* > 0$

$Y_1 = Yi^*$ if $Y_2 = 1$

Y_1 not observed if $Y_2 = 0$

X is a k-vector of regressors, Z is an m-vector of regressors, and the error terms U_1 and U_2 are jointly and normally distributed, independent of X and Z, with zero expectations. The first model is the one we are interested in. However, the latent variable Y_1 is only observed if $Y_2 > 0$.

3.3 Results

In the first part of our results, we provide an overview of some statistics regarding time allocation, activities performed per household member and individual earnings. This general part considers all household members including children, and all wives in polygamous households. In the second part related to earnings and factors of variation, in modelling individual performance per activity and across zones, we only consider the husbands and the first wives

3.3.1 Household production statistics

This section describes the sources of income earned from economic activity in the household by husband, wife and other adult members in rural households in the year of survey. It also estimates the variation of income according to zone, land availability, time used and other socio-economic factors. In rural households, economic activities are mainly related to agriculture. According to the United Nations, economically active persons are '*all the persons of either sex who furnish the supply of labour for the production of goods and services during the time-reference period chosen for the investigation*' (Blacker, 1978: 78). Following this definition, we investigated time spent by each household member, especially on agricultural activities. For non-agricultural activities (off-farm work), we report the income generated but we do not report the time use. This was impossible to evaluate because of the seasonality and irregularities of these activities. In addition, we report other income, mostly related to cash transfers from migrants.

Activity management and time use

In general, rural Senegalese households show a specialization of household members' daily activities. Results in Chapter 2 have shown that men, women and children tend to engage in specific agricultural activities in the household. The poor infrastructure for the provision of water and energy forces women and children more than men to allocate their time to acquiring these goods. Other female activities are milking and marketing milk in the market. Concerning livestock, males are more specialized in marketing, feeding and guarding ruminants.

The main source of income for the majority of the rural households is rain-fed farming (crop farming) and keeping livestock. Almost all households reported that farm work was their main activity. A detailed report concerning hours per day spent on farm activities shows that male household heads and active boys spent more time in crop activities than wives and active girls. On average, male heads spent 10 hours/day on livestock (water collection for the animal, feeding, guarding, etc.) and 8 hours/day on crops (shelling, cleaning, harvesting, marketing, etc.), compared to wives who spent on average 6 hours/day and 4 hours/day, respectively, on crops and livestock (wives also include those in polygamous households). For crops, the time reported is related to the rainy season from July to October, and we can see that males and females show almost the same performance (Table 3.1). This result is in line with other studies that suggest that in cash crops women are as productive as men (Saito *et al.*, 1994). Male heads also spent time (on average, 3 hours/day) in marketing ruminants in the market; this activity is not a concern of wives. In general and on average, active boys spend more time on livestock(12 hours a day) and crops (7 hours a day). Especially, boys' activities are guarding ruminants (on average, 5 hours/day) and cleaning fields and harvesting crops (Table 3.2). Other

Table 3.1. Time spent on activities by gender (average hours/day).

	Livestock		Food crops		Cash crops	
	Mean	Std	Mean	Std	Mean	Std
Male heads	10.42	0.66	3.32	0.20	5.28	0.21
Wives	4.11	0.35	2.76	0.32	3.55	0.28
Boys	12.38	0.72	3.75	0.25	3.68	0.25
Girls	1.44	0.16	1.24	0.11	1.53	0.14
Other dependents	3.25	0.36	0.22	0.08	0.26	0.08
Time spent by hired labour	3.22	0.82	1.46	0.22	1.75	0.32
Husbands marketing ruminants	3.17	0.18				

Table 3.2. Hours per day spent by children on agricultural activities.

Activities	Mean	Std
Boys guarding cattle	4.50	5.85
Boys guarding small ruminants	6.15	6.73
Boys watering	1.33	1.54
Boys clearing and harvesting	2.84	2.96
Boys shelling	0.82	2.15
Girls cleaning fields	1.22	1.92
Girls shelling	0.24	1.13
Girls milking	0.25	0.76
Girls watering	0.12	0.48

dependent members like helpers and relatives spend less time: 3 and 4 hours per day, on livestock and crops respectively.

An analysis by zone shows that male heads spent more time on livestock in the SP area (16 hours per day) than in the AP area (5 hours per day). In the SP area, much time was spent on men's guarding activities during the night (Table 3.3). The number of hours spent per day on crops and livestock seems to be high if we consider the aggregated time for the head of the household, but crop activity is only managed during six months in a year (May to October). In general, men spent less time in livestock activities during the rainy season where more time is devoted primarily to crop. For crops, the time spent averages 10 hours per day in the AP area and 7 hours per day in the SP area. These results showed the importance of livestock activities in the SP area and crop activities in the agro-pastoral area. The same can be observed for wives. Hired labour is especially reported in the AP area where time used in crops averages six hours per day. Active boys worked five times more in livestock in the SP area than in the AP area. Similar results have been observed in the AP area for crops. However, time spent by girls is marginal for crops in the SP area and for livestock in the AP area. Differences between areas are significant for all activities ($P < 0.001$) in terms of time allocation except for husbands in crop activities.

Table 3.4 shows that adult and young females are more involved in homework than males: cleaning, fetching water and wood, cooking, caring for children and ill persons. Indeed, it was quite difficult to estimate time related to these activities which represent an important part of household duties.

However, discussions with household members revealed that women have hardly any leisure time during the day. If females are not working in the fields or keeping livestock, they are occupied in the market (marketing milk or cereals) or performing

Table 3.3. Time used in farming by household members in the two different areas.

	Sylvo-pastoral area		Agro-pastoral area	
	Mean	Std	Mean	Std
Male heads in livestock	15.87***	1.01	5.08	0.60
Male heads in crops	7.25	4.06	10.26	0.55
Wives in livestock	7.38***	0.59	1.25	0.18
Wives in crops	2.24***	0.31	10.66	0.88
Boys in livestock	18.72***	1.03	5.92	0.72
Boys in crops	3.01***	0.45	11.85	0.66
Girls in livestock	2.37***	0.28	0.43	0.09
Girls in crops	1.02***	0.19	4.52	0.40
Other dependents in livestock	1.55***	0.45	4.33	0.56
Other dependents in crops	0.30***	0.12	7.07	1.00
Wives marketing ruminants	0.00	0.00	0.00	0.00
Male heads marketing livestock	5.63***	0.20	0.72	0.11

T-test (difference between areas): ***$P<0.001$.

Table 3.4. Housework involvement of household members (percentage).

	Repair/ fixing	Fetching water	Childcare	Cleaning/ cooking	Fetching wood	Care for ill persons
Male heads	79.26	12.04	30.33	20.74	37.79	70.33
Wives	37.46	85.67	94.31	89.50	58.67	72.00
Boys	40.27	23.08	6.67	14.11	41.28	8.33
Girls	21.81	59.20	38.80	59.36	36.45	23.83
Other	12.85	13.84	10.03	14.76	19.03	9.03

domestic duties. Males have more leisure time in the household. However, they are responsible for repairing and fixing things in the house (79% of respondents). Almost to the same extent as women (70% and 72%, respectively) they care for ill persons. In most African countries, empirical evidence found that women perform more homework than men (Brown and Haddad, 1995). Even if farming remains the most important source of income, the time spent on several other economic activities was not reported, although these may be a substantial source of income.

In fact, if we consider male heads of households in the two areas, we can see in Table 3.5 that 26% are involved in fattening cattle in the dry season, 22.6% are intermediary traders in livestock (buying and selling livestock for owners (*dioula*) and retailers (*tefanke*) and 6% earned money from other off-farm activities like crafts, home industries, masonry, or are employers in cities. Furthermore, 9.15% of all spouses in the household, 2.40% of young girls and 10% of adult boys also earned from other off-farm activities (especially as household helpers or gathering fodder for females' herds). A comparison by zone showed that households in the SP area are more involved in diversifying activities and earn more (Table 3.5).

This section has investigated household members' time used in agricultural activities and labour allocation in farming and in off-farm activities. The results brought out that male heads and young boys spent much time in livestock activities compared to women and young girls. Husbands and wives were almost equally involved in crop farming in terms of time allocated, especially in the AP area. While boys are mainly involved in guarding ruminants and cleaning fields and harvesting, young girls are mostly involved in homework and also in most agricultural activities (milking, shelling and clearing) especially in the AP area. Despite their contribution to farming, women are also very much engaged in housework activities (fetching water, childcare, clearing, fetching wood and caring for ill members). Moreover, they are also active in managing off-farm activities more than men (9.15% compared to 6.32%). A comparison between areas shows that SP farmers are mainly involved in livestock activities and AP farmers in crop growing activities. Also, households in the SP area are diversifying earnings more through off-farm activities than in the AP area.

Table 3.5. Household members' involvement in non-agricultural and dry-season activities (percentage).

	Sylvo-pastoral area	Agro-pastoral area	Total sample
Males fattening cattle	25.30	27.15	26.22
Males as animal traders	40.25**	4.60	22.42
Males in off-farm activities	4.00	8.64	6.32
Wives in off-farm activities (all wives)	16.30	2.00	9.15
Wives in off-farm activities (monogamous households)	25.16**	1.32	13.24
Adult boys in off-farm activities	8.75	11.25	10.00
Adult girls in off-farm activities	2.15	2.65	2.40

T-test (difference between areas):** $P<0.01$.

Household income from farming

To study income procurement, gross income has been considered and evaluated for crop-growing activities (food crops, cash crops, and garden vegetables), livestock (selling goats, sheep and cattle, fattening and marketing milk), trade (all goods), off-farm income during the dry season and other money transfers. In fact the crop-growing activities are managed during the rainy season which starts generally in July and ends in September or October. Fattening livestock is mainly practiced during the dry season from March to June in order to generate earnings to meet household needs. This analysis is limited to the heads of the households and dependent or subordinate members. In polygamous households (which represent 38% of the sample), the average over all wives was considered.

A first analysis concerning only the income from livestock and crops appears to show that the husband's income from farming is the most important part of the household income. Women achieved lower crop and livestock output than men (Table 3.6). Husbands' income from farming represents almost 81% of the total household income from farming, while wives' income from crops and livestock amounted on average to 15.45% of the total household earnings from farm activities. Concerning children, adult boys earned more than adult girls in farm activities, 3.12% and 0.39% of household income, respectively. On average, other household members (like dependent helpers) have contributed 0.71% of total household income from farm activities. A comparison by zones showed that for male heads, herding is more important in the sylvo-pastoral area than in the agro-pastoral area. In contrast, earnings related to crops in the AP area total almost ten times those in the SP area. Regarding females, the same result has been noticed: females in the SP earn more from livestock, and those in the AP area earn more from crop growing. Apart from their own earnings from farm and off-farm activities, women also received compensation from their husbands for their work in cash crops production and from selling livestock. This monetary compensation is more visible in the AP area (Table 3.6). However, the monetary compensation from husband to wife represents only 7.90% of women's earnings and cannot explain entirely the economic contribution of women in agricultural activities. The largest part of women's earnings is determined from livestock production (milk production and livestock assets).

An independent sample T-test between both areas (estimated only for the shaded items in Table 3.6), for total household earnings related to farm activities (including crops and livestock), shows a significant difference between the AP area and the SP area ($P < 0.05$). However, the difference between the two areas in earnings for male heads is not significant. In contrast, wives earn significantly more farm income in the SP area than in the AP area ($P < 0.01$). These results suggest that total farm income is variable across areas and that women in the SP area seemed to be wealthier, economically speaking, than females in the AP area.

Table 3.6. Average income in CFA[1] of household members for crops and livestock in the survey year.

	Total (N=300)		SPA (N=149)		APA (N=151)	
	Mean	Std	Mean	Std	Mean	Std
Husbands' income from crops	266,867	32,625	43,309	10,921	487,463	58,702
Wives' income from crops	7,730	1,638	2,631	16,591	12,761	29,382
Husbands' income from livestock	1,079,200	87,862	1,404,200	135,596	764,934	106,955
Wives' income from livestock	255,738	31,102	437,285	57,465	76,596	13,609
Wives' compensation from husbands	26,218	66,237	24,201	83,652	28,194	43,130
Total husbands' income from trade	378,635	58,482	478,165	106,942	283,059	50,182
Total husbands' income from crops and livestock	1,724,400	178,970	1,925,674	274,647	1,535,456	215,838
Total wives' income in trade	41,907	10,480	39,054	20,185	44,677	7,454
Total wives' income' from crops and livestock	331,593	43,220	503,171	94,241	162,328**	23,975
Adult boys' income from crops	5,266	1,910	2,584	1,833	7,914	3,329
Adult boys' income from Livestock	61,775	10,879	102,231	18,928	21,854	9,926
Adult girls' income from crops	50	866	-	-	99	99
Adult girls' income from livestock	8,416	2,417	16,946	4,775	-	-
Total other members' income	15,416	5,818	17,718	9,308	13,146	7,046
Total household income from crops and livestock	2,146,916	182,324	2,568,324	304,105	1,740,797*	192,321

T-test (difference between areas; only estimated for the figures shaded in grey): *$P<0.05$, **$P<0.01$.

[1] 1 US$ = 475 CFA during the year of the survey.

Further analysis shows that male heads of the household are more productive in crop production than females; they earn 247.00 CFA per hour (if we consider 6 months of working hours in all crop activities), while women earn 7.15 CFA per hour for almost the same time spent on crops (six hours for wives and 8 hours for husbands). This result suggests that males gain 35 times more than females for the same number of hours spent on crops. This might be explained by the fact that men sell all household crop produced and keep the revenues for themselves, especially from cash crops (see also Rowling, 2008). In livestock production males gain 300 CFA per hour compared to 175 CFA for females; even though statistically significant, the difference is far less than for crop income, suggesting that in livestock production, women are less disfavoured than men. This result is explained by the fact that women are responsible for selling milk at the market and retain their earnings. In line with our results, findings from the Peruvian Sierra (Jacoby, 1991) also suggested that female labour is less productive in crop than in livestock activities. Analysis across areas shows that male heads are more productive in the AP area than in the SP area for both crop and livestock activities (375 CFA per hour in the AP area compared to 40 CFA in the SP area for crops, and 420 CFA per hour in the AP area compared to 242 CFA in the SP area for livestock). In fact, farmers in the AP area benefit from the advantages of mixed farming systems (where agriculture and livestock are complementary and mutually beneficial) and better prices because of the proximity to big cities. For females, productivity (earnings over time allocation to crop and livestock separately) does not show much difference between the two areas.

Off-farm household income base

In the entire sample, total off-farm activities accounted for 5.14% of the total household income and the largest share was earned by male heads of households (on average 38.94%). Wives and other dependents achieved respectively 27.47%, 30.33% (adult boys), 2.21% (adult girls) and 1.03% (other). Data showed large differences between zones (Table 3.7), especially regarding husbands' and wives' off-farm wages. However, differences in earnings were only significant for wives' earnings ($P < 0.001$) as in our previous analysis concerning farm activities; off-farm wages were much higher in the SP area than in the AP area. Cash transfers to the households (especially from

Table 3.7. Average off-farm income by household members in the year of the survey (average).

	Global		SPA		APA	
	Mean	Std	Mean	Std	Mean	Std
Total male heads	46,869	33,822	70,777	67,580	23,280	8,770
Total wives	33,064	6,188	59,863	10,857	7,152**	5,421
Total boys	36,505	9,320	34,831	14,989	38,146	11,218
Total girls	2,667	1,365	1,812	1,109	3,510	2,486
Others	1,246	726	822	821	1,656	1,190
Total	120,351	61,908	168,105	120,089	73,744	18,500
Other transfers	71,581	17,200	64,336	27,487	78,444	21,147
Total households' non-agricultural income	191,932	401,410	232,441	483,247	152,188	310,551

T-test (difference between areas, shaded in grey to emphasize difference): $^{**}P<0.01$.

relatives abroad) accounted for 37.30% of total off-farm income and 3.06% of total household earnings.

The results in Tables 3.6 and 3.7 regarding earnings in farming activities and off-farm activities show that there is a significant difference between women's earnings across zones. Women in the SP area earn more than in the AP area from farming and off-farm activities. The difference is large and may suggest that women in the SP area are not constrained in keeping earnings for themselves and engaging in activities outside the household. The difference in earnings between males and females is greater in the SP area than in the AP area. However, such difference may derive from the fact that farmers in the SP area are wealthier and gain more income from livestock.

3.3.2 Household income determinants

In this section we consider whether human capital (education, age, labour, land use, time) and social, cultural and demographic factors like ethnicity (people's social membership and geographic location) have an impact on total earnings for two activities: livestock and crops in the two areas. We assume that off-farm earnings or market wages which represent 5.14% of total household income, are important as has been shown in previous results. For this reason, non-agricultural income was included as explanatory variable in the economic framework of household production. Since women as well as men are involved in all farming activities (cash crops, food crops and livestock), we break down all explanatory variables by gender so that factors associated with gender earnings are explained more clearly. This approach takes into consideration an important criticism underlined by some authors (e.g. Quisumbing, 1996) in analyzing sex differences in technical efficiency. The criticism concerned the estimation of only one equation and using a dummy variable for gender, instead of estimating separate equations for males and females.

The household income correlations have been identified for crop and livestock activities using Heckman's two-step selection model. The model follows two stages: stage one, referred to as the selection equation, is estimated by a probit model on the entire data set to capture the determinants of censoring. The variables included at this stage do not comprise all variables included in Equations 3.3 and 3.5. The variables in the selection equation are restricted only to those that can affect the censoring, including human capital variables like education, ethnicity, area, no agricultural income (transfers and off-farm earnings) and ownership of livestock. We also expect that education, represented here by different levels of formal education (primary school and secondary school) and informal education (basic reading and writing), will affect the censoring. Stage 2 is related to the income equation concerning the non-censored observations. This stage refers to an OLS on valid observations. The variables included at this stage may refer to all variables included in Equation 3.3.

As a measure of experience, we include age and age squared, of females and male heads who are most appropriate to capture the effect of years in farming. Also, education and ethnicity may be taken into account at this stage. The test for the joint null hypothesis that the coefficients of the variables common to both Heckman steps are equal, is strongly rejected in the SP area at $P < 0.01$ for crops and for the livestock equation the test is rejected at $P < 0.01$. In the AP area, the test of the joint null hypothesis is rejected at $P < 0.01$ for crops and $P < 0.05$ for livestock.

Livestock

The important positive factors that are associated with livestock production in the AP area (see Table 3.8) are as expected the number of cattle ($P < 0.001$), involvement of girls and husbands in caring for ill members ($P < 0.001$) and husbands' education ($P < 0.01$). For the number of cattle, 1% increase in the number leads to a 0.42% rise in earnings from livestock. These results are in line with the work of Fafchamps and Quisumbing (1998) in Pakistan who found positive income correlation of livestock production. In addition, cultivated land for millet and water melon is a significant predictor of the participation in livestock (respectively at $P < 0.05$ and $P < 0.01$). The girls' involvement in caring for ill household members has a negative effect on crop earnings in terms of correlation ($P < 0.01$). This can be explained by the fact that girls are productive in housework (cleaning and fetching wood and water), so their caring for ill members may allow both husband and wife to be less productive in livestock tasks, thus inducing a lower income. It has been argued that girls' labour supply is a substitute for wives' labour supply: the more involved wives are in farming, the less girls are involved in farming and the more they are involved in housework (see also Grootaert and Patinos, 1999). The same effect is noticed for husbands: their caring for ill persons negatively affects livestock income ($P < 0.001$). In fact, caring, which is generally done by wives, can potentially affect welfare and well-being in the long term. For example, Newell (1993) found that caring has a positive effect on household productivity. Moreover, males could compensate for the burden of caring by reducing their leisure time but not their working time (Gultierez, 1998). However, in the short run, caring may be negatively associated with earnings. Other findings report that through an indirect effect (e.g. World Bank, 1999), more male heads' involvement in caring could increase the propensity of females to participate in farming activities, thus increasing gross income from livestock.

The positive association between the husband's education and livestock production suggests that more knowledge can enhance better use of technologies like cross-breeding which enables higher milk production and thus more income for the household. The bad health of females is negatively correlated with earnings from livestock ($P < 0.05$). Health problems lead to 1.84% decrease on earnings from livestock. Distance from the market does not show a significant effect.

Table 3.8. Heckman's two step selection model; dependent variable household income from livestock in the AP area.

	Coefficient	Std	t-value
Outcome equation			
Wolof group [6]	0.5040	0.3383	1.49
Peulh [7]	0.8063	0.4062	2.03*
Husband's age (log)	0.1246	0.0738	1.69
Wife's age (log)	0.0015	0.0727	0.02
Husband's education [1]	0.2099	0.0819	2.43**
Wife's education [1]	-0.1461	0.1026	-1.52
Wife's access to land [2]	-0.1487	0.2715	-0.55
Land cultivated (log)	-0.1543	0.0968	-1.59
Household size (log)	-0.4799	0.2763	-1.74
Number of cattle (log)	0.4261	0.1060	4.06***
Number of goat (log)	0.0610	0.0907	0.67
Number of sheep (log)	0.0885	0.0874	1.01
Time marketing animals (H)	0.2391	0.2134	1.12
Time in livestock in log (H)	0.0434	0.0991	0.44
Time in livestock in log (W)	-0.0453	0.1741	-0.26
Time in livestock in log (B)	-0.0004	0.0895	0.00
Time in livestock in log (G)	0.1773	0.2176	0.81
Time in livestock in log (Helpers)	0.0439	0.0977	0.45
Housework (H) [4]	0.1576	0.1324	1.19
Housework (W) [4]	-0.0707	0.1154	-0.61
Housework (B) [4]	-0.0297	0.1230	-0.24
Househork (girls) [4]	0.1341	0.0786	1.71
Househork (Helpers) [4]	-0.0449	0.0659	-0.68
Husband's caring ill members [4]	-1.1627	0.3460	-3.37**
Wife's caring ill members [4]	0.1375	0.2441	0.56
Boy's caring ill members [4]	1.4435	0.4599	2.87**
Girls caring ill members [4]	-0.8748	0.3417	-2.67**
Off-farm earnings (log)	-0.0089	0.0180	-0.50
Health status (H) [5]	-1.0103	0.6578	-1.55
Health status (W) [5]	-1.8440	0.9578	-1.95*
Distance to market (log)	0.1015	0.1702	0.60
Cash transfers (log)	0.0069	0.0251	0.27
Constant	10.3449	1.6219	6.61

Table 3.8. Continued.

	Coefficient	Std	t-value
Selection equation			
Wolof group [6]	0.0018	0.3793	0.00
Peulh [7]	0.4271	0.5839	0.73
Peanut (yield in log)	-1.0263	0.4575	-2.25*
Water melon (yield in log)	1.3527	0.5112	2.61**
Maize (yield in log)	0.4851	0.3771	1.29
Millet (yield in log)	1.0427	0.4624	2.28*
Beans (yield in log)	0.3653	0.5531	0.66
Wife's savings (log)	-0.0006	0.0290	-0.02
Wife's access to land	0.3520	0.5546	0.63
Off-farm earnings (log)	-0.0339	0.0288	-1.18
Cash transfers from migrants (log)	-0.0491	0.0298	-1.64
Constant	0.3276	0.6861	0.48
Lambda	-0.8475	0.4834	-1.75
Rho	-0.9303		
Sigma	0.9109		
N=133			
Censored=26			
Uncensored=107			
Wald test $chi^2(33)=145.11$			
Prob $chi^2<0.0001$			

$^*P<0.05$, $^{**}P<0.01$, $^{***}P<0.001$.

H = husband, W = wife, B = boy, G = girl, O = other dependent members.

[1] Education dummy:1 = highest education.

[2] Dummy wife's access to land: 1= yes.

[3] Dummy wife's access to credit: 1= yes.

[4] Due to difficulties in reporting hours in housework by farmers and their children, we prefer using their involvement and take housework as dummy: 1 = yes.

[5] Dummy: 0 = illness problem, 1 = good health.

[6] Dummy Wolof (Wolof=1, Sereer=0).

[7] DummyPeulh (Peulh =1, Sereer=0).

In the SP area, there are no censored observations because 98.7% of the sample participates in livestock activities. The regression of Equation 3.3 is estimated by OLS on the valid observations yielding an adjusted R^2 of 58% (Table 3.9). Crop type is correlated differently with income from livestock in the SP area than in the AP area. While millet has a positive correlation with income ($P < 0.001$), cultivated beans decrease livestock income. This is due to the fact that millet is mainly used for the own household's consumption and bean is a cash crop and competes with earnings from livestock. The husband's time spent on marketing ruminants has a positive association with livestock income ($P < 0.001$): 1% increase in time spent on marketing ruminants leads to 1.35% increase in total livestock income. In the SP area, the wife's time allocated to caring for ill members has a positive correlation with livestock income. For boys, the same effect is noticed for housework (fetching wood and water and clearing). Regarding human capital, the husband's age has a positive association with livestock in the SP area ($P < 0.001$), suggesting that experience and knowledge acquired over the years is a productive factor of livestock farming. Cash transfers in the SP area have a negative effect ($P < 0.001$) on livestock income, leading to a 0.20% decrease for each additional CFA (as a percentage) of cash transfers. This effect is probably due to more investment in crops in order to become more sedentary, and to the redistribution of labour force participation (the household diversifies activities when cash transfers are available). The results highlight the fact that alternatives which enable the generation of cash income like transfers from abroad or from relatives in other cities, do not favour participation in livestock production. Moreover, since farming is highly dependent on climate (in Senegal, the rainy season lasts only three months) and has to cope with constraints or difficulties arising from low production systems, the cash income from migration is highly welcome (e.g. Rosenzweig and Stark, 1989; Fall and Sy, 2004). Cash transfers from migrants can also be used to fulfil other basic household needs (housing, building toilets). In line with our findings, Alderman and Chishti (1991) found that income from cash transfers reduces labour force participation in the farming system. As for housework, involvement of both women and boys have a positive correlation with earnings ($P < 0.01$). For women, the effect is noticed in caring for ill members. The distance from the market and the health status do not show significant effects.

After computation of the beta coefficients, we found that the positive effect of husband's time allocated to marketing ruminants is economically meaningful; a change of one standard deviation in quantity will result in a 0.29 standard deviation change in household earnings from livestock. Also, the significant positive effect of wife's caring for ill members is economically meaningful (one standard deviation change leads to 0.21 standard deviation change in livestock earnings). Looking into the beta coefficient method allows us also to make comparisons of the relative importance of the independent variables. It turns out that cash transfers from migrants appear to have an economically significant effect on livestock earnings (leads to 0.25 standard deviation change in livestock earnings). Other variables such as women's

Table 3.9. Regression analysis: dependent variable: household income from livestock in the SP area (OLS).

	Coefficient	Std	t-value
Husband's age (log)	0.4687	0.1269	3.59***
Wife's age (log)	-0.2267	0.1176	-1.93*
Husband's education [1]	-0.1222	0.1697	-0.72
Wife's education [1]	-0.3176	0.3208	-0.96
Wife's access to land [2]	-0.0149	0.4861	-0.03
Peanut (yield in log)	-0.6963	0.4148	-1.68
Water melon (yield in log)	-0.8677	0.6202	-1.40
Maize (yield in log)	0.3155	1.3517	0.23
Millet (yield in log)	1.0684	0.3607	2.94**
Beans (yield in log)	-1.3129	0.5987	-2.19*
Household size (log)	0.2113	0.5135	0.41
Number of cattle (log)	-0.2299	0.2071	-1.11
Number of goat (log)	-0.3283	0.1921	-1.71
Number of sheep (log)	0.2184	0.1942	1.12
Number of horses and donkeys	0.0653	0.1848	0.35
Distance to market (log)	-0.0539	0.1727	-0.31
Time marketing animals (H)	1.3520	0.4155	3.22**
Time marketing animals (B)	0.2184	0.3179	0.69
Time in livestock in log (H)	0.2366	0.1560	1.52
Time in livestock in log (W)	-0.0282	0.2645	-0.11
Time in livestock in log (B)	-0.2029	0.2330	-0.87
Time in livestock in log (G)	-0.1044	0.2363	-0.44
Time in livestock in log (Helpers)	-0.0604	0.3900	-0.15
Health status (H) [3]	-0.6936	0.6156	-1.13
Health status (W) [3]	-0.0915	0.7283	-0.130
Housework (H) [4]	-0.2183	0.1592	-1.37
Housework (W) [4]	-0.2182	0.2058	-1.06
Housework (B) [4]	0.7925	0.2433	3.24**
Househork (girls) [4]	-0.2090	0.1357	-1.54
Househork (Helpers) [4]	-0.7362	0.8842	-0.83
Husband's caring ill members [4]	-0.3813	0.5387	-0.71
Wife's caring ill members [4]	2.0288	0.5411	3.56**
Boy's caring ill members [4]	-0.8255	0.6044	-1.37
Girls caring ill members [4]	-0.3405	0.5897	-0.58
Wife's savings (log)	0.0249	0.0371	0.67
Off-farm earnings (log)	-0.0008	0.0300	-0.03
Cash transfers from migrants (log)	-0.2018	0.0488	-4.08***
Constant	3.8838	3.2946	1.18

Table 3.9. Continued.

	Coefficient	Std	t-value
N=105			
Adjusted R^2=0.58			

$^*P<0.05$, $^{**}P<0.01$, $^{***}P<0.001$.

H = husband, W = wife, B = boy, G = girl, O = other dependent members.

[1] Education dummy: 0= lowest, 1 = highest education.

[2] Dummy wife's access to land: 1= yes.

[3] Dummy: 0 = illness problem, 1 = good health.

[4] Due to difficulties in reporting hours in housework by farmers and their children, we prefer using their involvement and take housework as dummy: 1 = yes.

involvement in chores do not induce large changes in livestock earnings (a change of one standard deviation in time allocated to housework will only result in 0.042 standard deviation change in livestock earnings in the SP area).

Crop activities

In the AP area the number of sheep decreases household income from crops ($P<0.05$) (Table 3.10), but the number of cattle is a positive factor of participation ($P<0.05$). This may be explained by the fact that keeping small ruminants requires more space which is not available in the AP area, so additional goats and sheep have a depressing effect on yield cultivated, thus decreasing crop income. On the other hand, keeping cattle generates more revenues through milk production and more natural fertilizers which can be used for crops. These results show that the composition of herds (the kinds of animals involved) competes with growing crops: farmers who have more sheep, earn less on crops, and those who own more cattle are more involved in growing crops. The major crops cultivated also increase crop income considerably (water melon ($P<0.01$), maize ($P<0.01$), millet ($P<0.01$) and groundnut ($P<0.001$)). The wife's access to credit is a negative factor in crop income ($P<0.001$), suggesting that access to credit pushes wives away from crop activities, probably through an investment of credit in other activities like trading or keeping livestock (which enables better milk production, milk processing and marketing). In fact, access to credit (in this case micro-credit) allows the women to get involved in productive activities like self-employment and entrepreneurial activities that are physically less demanding than agricultural activities. These activities enable more savings and increase women's autonomy and self-confidence, providing opportunities for poverty alleviation (see also Nowak, 2008). The wife's education has the same negative correlation with her participation in growing crops.

Table 3.10. Heckman's two-step selection model; dependent variable: household income from crops in the AP area.

	Coefficient	Std	t-value
Outcome equation			
Age (H)	0.0095	0.0096	0.99
Age (W)	-0.0076	0.0097	-0.78
Education (H) [1]	0.0056	0.0507	0.11
Education (W) [1]	-0.2185	0.0861	-2.54**
Wife access land [2]	-0.2425	0.1695	-1.43
water melon (log)	0.2659	0.1098	2.69***
Peanut (log)	0.5477	0.1618	4.00**
Maize (log)	0.3235	0.1197	3.05***
Millet (yield in log)	0.5174	0.1869	3.22***
Bean (yield in log)	-0.0570	0.2029	-0.28
Access to credit [3]	-0.5269	0.1335	-4.34**
Household size (log)	0.0867	0.2058	0.42
Goat number (log)	0.0040	0.0614	0.07
Sheep number (log)	-0.1567	0.0895	-2.20**
Cattle number (log)	0.0834	0.0564	1.48
Time in crops in log (H)	0.0453	0.0812	-0.56
Time women in crop (log)	0.0492	0.0485	1.02
Time in crop in log (B)	-0.2441	0.0945	-2.58**
Time on crop in log (O)	0.0218	0.0470	0.46
Housework (H) [4]	0.0238	0.0650	0.37
Housework (W) [4]	-0.1450	0.0698	-2.10*
Housework (B) [4]	0.0094	0.0808	0.12
Housework (G) [4]	0.0328	0.0432	0.76
Housework (O) [4]	-0.1028	0.0307	-3.31**
Distance to market (in log)	-0.0082	0.1027	-0.08
Health status (H) [5]	-0.2022	0.7176	-0.28
Health status (W) [5]	0.0334	0.3169	0.11
Constant	11.64604	1.0289	11.32
Selection equation			
Wife 's access to land	-0.2147	0.3950	-0.54
Off-farm earnings (log)	-0.0296	0.0287	-1.03
Transfers (log)	0.0056	0.0359	0.16
Wife's savings (in log)	-0.0130	0.0307	-0.42
Education (H)	-0.1002	0.1240	-0.81
Education (W)	-0.3415	0.1627	-2.42**
Goat number (log)	0.0891	0.1505	0.59

Table 3.10. Continued.

	Coefficient	Std	t-value
Selection equation (continued)			
Sheep number (log)	-0.3041	0.1413	-1.90**
Cattle number (log)	-0.2713	0.1287	-2.11**
Constant	2.0464	0.7237	3.67
Lambda	0.2884	0.2645	0.19
Rho	0.6123		
Sigma	0.4718		
N=133			
Censored observation=27			
Uncensored=106			
Wald $chi^2(31)=412.25$			
Prob $chi^2<0.0001$			

*$P<0.05$, **$P<0.01$, ***$P<0.001$.

H = husband, W = wife, B = boy, G = girl, O = other dependent members.

[1] Education dummy: 0= lowest, 1 = highest education.

[2] Dummy wife's access to land: 1= yes.

[3] Dummy wife's access to credit: 1= yes.

[4] Due to difficulties in reporting hours in housework by farmers and their children, we prefer using their involvement and take housework as dummy: 1 = yes.

[5] Dummy: 0 = illness problem, 1 = good health.

Regarding time allocation in the AP area, we found that the involvement of other dependent members (relatives) in housework is negatively associated with earnings.- probably because they are more productive in other activities. This result suggests that the involvement of other dependent members in household chores reduces the labour supply to crop activities, resulting in less productiion. Women's and other members' involvement in housework also are negatively correlated with total total earnings from crops (respectively at $P<0.05$ and $P<0.01$). This result suggests that their participation in housework induces less labour supply to crops thus decreasing earnings. With respect to the participation in crop activities, we found that better-educated wives are less probably involved in farming ($P<0.01$). This finding suggests that such wives push women out of growing crops, probably through their involvement in other activities. In the AP area neither health status nor distance from market has any association with crop earnings.

Because we lose observations due to censoring (78.26%), in the SP area the number of variables in the outcome equation becomes too large compared to the number of observations. For this reason, we restrict the number of variables in the outcome equation by aggregating some of them (time allocation). Wives' access to land has a positive effect on crop earnings in the SP area (Table 3.11). This finding may be explained by the fact that women with access to land mainly grow food crops (millet, bean) mostly for their own household consumption, thus decreasing the costs of food consumption. Because labour force participation is an important factor of production, household size predictably is associated positively with earnings from crops ($P<0.01$). With respect to time allocation, a negative correlation on crop earnings was found for wives. This result may be explained by the fact that growing crops in the SP area is not productive. Because women and boys are productive in other activities, their involvement in housework also is negatively correlated with crop earnings. Distance from the market negatively affects crop income in the SP area ($P<0.001$). One percent increase in distance leads to a 0.41% decrease in earnings from crops.

While in the SP area the husband's education is negatively correlated with crops ($P<0.05$), in the AP area the wife's education has a negative effect on the probability of participation in growing crops ($P<0.01$). However, better-educated husbands show involvement in off-farm work and consequently higher yearly off-farm wages (thus applying income diversification strategies) (Table 3.12). Most studies in rural farming systems found that education could be negatively and significantly associated with participation in income-generating activities and in most cases, with an increase in leisure time (Ilahi and Jafarey, 1999; Alderman and Chishti, 1991; Ilahi, 2000).

Also, while cash transfers from migrants are negatively associated with earnings from livestock activities in the SP area, they allow farmers there to be more involved in crops. In addition, distance from the market decreases earnings from crops ($P<0.001$).

In conclusion, the results show that there are important differences between the agro-pastoral area and the sylvo-pastoral area. In mixed farming systems (mainly in the AP area), where household members are more involved in farming activities, the combination of growing crops and keeping livestock is influenced by important variables like household size, the husband's and wife's education and also ethnicity. The interdependency between livestock and crop activities is highly visible in the AP area because the number of cattle is an important determinant of livestock earnings and crop growing. In the SP area, cash transfers from migrants show important but different consequences of growing crops and keeping livestock. In fact, cash transfers allow for more investment in crops and less involvement in livestock, suggesting that with alternative revenues, pastoralists tend to become more sedentary through growing more crops. In our sample, 35% of the sample in the SP area received cash transfers compared to 30% in the AP area. These results also suggest that gender differently affects the activities in the two areas. The

Table 3.11. Heckman's two step selection model; dependent variable: household income from crops in the SP area (in log).

	Coefficient	Std	t-value
Outcome equation			
Husband's education[1]	-0.2117	0.1438	-2.09*
Wife's education[1]	-0.3512	0.2515	-1.40
Wife's access to land[2]	0.9230	0.4409	2.09*
Land cultivated in log	0.0911	0.2236	0.41
Wife's access to credit[3]	-0.1908	0.3143	-0.61
Household size (log)	0.3922	0.3960	2.86**
Number of goat (log)	0.1280	0.1621	3.36***
Number of sheep (log)	-0.5539	0.1515	-4.61***
Number of cattle (log)	-0.0025	0.1359	-0.02
Time spent in crops by other (H, O, B, G)	-0.1746	0.2500	-0.59
Time spent in crops (W)	-0.3131	0.1114	-2.75**
Housework (H) [4]	-0.2122	0.1404	-1.51
Housework (W) [4]	-0.7849	0.1421	-5.38***
Housework (B) [4]	-0.5236	0.1016	-5.156***
Housework (G) [4]	0.4227	0.8311	5.08***
Health status (H) [4]	-0.8273	0.4861	-1.70
Distance (log)	-0.4214	0.0804	-5.24***
Constant	16.4007	1.5990	10.56
Selection equation			
Wife's access to credit	0.4838	0.4042	1.20
Land cultivated (in log)	0.1718	0.1328	1.29
Off-farm earnings (log)	0.0459	0.0312	1.47
Cash transfers from Migrants (log)	0.1459	0.0432	3.38**
Wife's savings (log)	-0.0310	0.0454	-0.68
Husband's education	-0.0473	0.1490	-0.32
Wife's education	-0.0809	0.2491	-0.32
Constant	-2.0451	0.6409	-3.19
Lambda	0.3228	0.3047	1.32
Rho	0.7420		
Sigma	0.5729		
N=115			
Censored observations=90			
Uncensored observations=25			
Wald $chi^2(15)=63$			
Prob $chi^2<0.0001$			

Table 3.11. Continued.

$^{*}P<0.05$, $^{**}P<0.01$, $^{***}P<0.001$.
H=husband, W= wife, B=boy, G=girl, O=other dependent members.
[1] Education dummy: 0= lowest, 1 =highest education.
[2] Dummy wife access to land: 1=yes.
[3] Dummy wife's access to credit: 1=yes.
[4] Due to difficulties in reporting hours in house work by farmers and their children, we rather use their involvement and take housework as dummy: 1=yes.

Table 3.12. Comparison between yearly off-farm earnings and husbands' level of education (in CFA)[1].

Levels of education	Mean	Std
Illiterate and basic level (N=243)	327,506	74,000
Highest level (Primary school or Secondary level) (N=46)	786,425	333,575

[1] 1 US$ = 475 CFA during the year of the survey

involvement of girls in homework (fetching wood, water and cleaning) in the AP area makes women more productive in other activities. Consequently, women's time allocated to household tasks reduces income from crop growing. But time allocated to household tasks (fetching wood, cleaning and childcare) carried out by other dependent members, is also a negative factor in generating income from crops. The economic effects of some variables such as husband's time allocated to marketing animals, women's involvement in caring for ill members and housework and also cash transfers from migrants are meaningful because the beta coefficients are relatively large (exceeding 0.30). Regarding education, better-educated male heads show higher earnings from livestock in the AP area, while in the SP area the association with crop is negative. For women, higher education negatively affects the probability of participation in crop activities, suggesting that better-educated wives are less interested in farming. In the same area (SP), distance from the market also has a negative correlation with earnings while health problems are negatively associated with earnings from livestock.

3.3.3 Individual income determinants

In the estimation of the Heckman regressions, total income as the dependent variable for crops and livestock is considered separately for female and male household heads. Considering the individual level, we expect the following variables to have direct effects on their total earnings for each activity, x. (1): time spent on farming

(T_{xi}), which can be divided into time spent on crop activities and time spent on livestock activities, (2) involvement in homework (H_{xi}) which can be divided into several variables (caring for ill members and children and other chores), (3) land availability for different crops (L_{dxi}), (4) owning cattle, sheep and goats (L_{xi}) and other characteristics of the household (females' household goods considered as assets, wife's savings, wife's access to credit, area, distance from the market and health status). Education and ethnicity have been considered as dummy variables. The equation model can be specified as:

$$Y_{xi} = \alpha_0 \, T_{xi}^{\alpha_1} \, H_{xi}^{\alpha_2} \, L_{dxi}^{\alpha_3} \, L_{vxi}^{\alpha_4} \, e^{\alpha_5 z} \, e^{u} \qquad (3.8)$$

After taking the logarithm, this can be written as:

$$log\,(1+Y_{xi}) = log\,\alpha_0 + \alpha_1 log\,(1+T_{xi}) + \alpha_2 log(1+H_{xi}) + \alpha_3 log(1+L_{dxi}) + \alpha_4 log(1+L_{vxi}) + \alpha_5 Z + u \qquad (3.9)$$ [5]

Individual income (Equation 3.8) is a censored variable since 14.16% of the male heads in the sample have no income from livestock and 40.16% have no income from crops. For female heads, censored observations are noticed for livestock (39.66%) and also for crops (91%). For this reason, we use Heckman's two-step model in order to consider all observations and to allow for participation in farming to be determined by variables different from those that determine the value taken by the positive observations. Variables are also allowed to differ in the two equations, and to take different coefficients. The test for the joint null hypothesis that the coefficients of the variables common to both decisions are equal in the two equations, is rejected at $P<0.01$ in the male crop equation, at $P<0.01$ in the male livestock equation, at $P<0.10$ in the wife's livestock equation and at $P<0.05$ in the wife's crop equation, showing the robustness of the model.

Determinants of wife's income from farming

Determinants of wives' participation in crop activities are their savings and the number of cattle (Table 3.13). This number decreases their probability to participate in crops activities ($P<0.01$), because of the competition between the activities in terms of time allocation. Wife's savings also are negatively associated with the participation in crops ($P<0.05$), suggesting that savings enable women to get involved in other activities (trade and small entrepreneurship). In the outcome equation, results show that household size positively affects wife's income from crops. The same is noticed for off-farm activities which positively affect earnings from crops ($P<0.05$). The effect of off-farm earnings may be related to the use of better seeds and inputs in plots and also to the access to new technologies (machinery). An important positive

[5] We add 1 to the explanatory variables in order to avoid computational problems when taking the logarithm.

Table 3.13. Heckman's two step selection model; dependent variable: wife's income from crops (in log).

	Coefficient	Std	t-value
Outcome equation			
Husband's education [1]	-0.3948	0.1854	-2.13*
Wife's education [1]	0.4423	0.2125	2.08*
Wife's access to land [2]	-0.0406	0.4079	-0.10
Household size (log)	0.9600	0.3526	2.72**
Time in livestock in log (W)	-1.1401	0.1947	-5.86***
Time spent in crops in log (W)	0.6670	0.4804	1.39
Zone [6]	0.7391	1.0195	0.72
Time spent in crops by other members (in log)	-0.0689	0.6358	-0.11
Wife's off-farm earnings	0.2195	0.0991	2.21*
Wife's access to credit [3]	-0.2071	0.1741	-1.19
Land cultivated (in log)	-0.2819	0.2398	-1.18
Housework (H) [4]	0.0493	0.0905	0.54
Housework (B) [4]	-0.0586	0.1056	-0.55
Housework (W) [4]	0.0843	0.0634	1.33
Housework (G) [4]	0.1311	0.0636	2.06*
Housework (O) [4]	0.0448	0.0313	1.43
Distance from markets (in log)	-0.7013	0.1478	-4.74***
Health status (W) [5]	0.0552	0.3251	0.17
Constant	5.1822	2.2800	2.27
Selection equation			
Husband's education	0.0979	0.1139	0.86
Wife's education	-0.3491	0.2186	-1.60
Wife's access to land	-0.4015	0.4033	-1.00
Cattle (number in log)	-0.4637	0.1943	-2.39**
Sheep (number in log)	0.1327	0.1472	0.90
Land cultivated (log)	0.2656	0.1284	2.07*
Off-farm earnings	0.0590	0.0364	1.62
Wife's access to credit	0.0694	0.3127	0.22
Wife's savings (log)	-0.0558	0.0277	-2.02**
Constant	-0.8053	0.6221	-1.29
Lambda	0.0169	0.3933	0.04
Rho	0.1066		
Sigma	0.1583		

Table 3.13. Continued.

	Coefficient	Std	t-value
N=264			
Censored observations=240			
Uncensored=2			
Wald chi^2(18)=267.29			
Prob chi^2<0.0001			

**P*<0.05, ***P*<0.01, ****P*<0.001.
H = husband, W = wife, B = boy, G = girl, O = other dependent members.
[1] Education dummy: 0= lowest, 1 = highest education.
[2] Dummy wife access to land: 1 = yes.
[3] Dummy wife's access to credit: 1= yes.
[4] Due to difficulties in reporting hours in house work by farmers and their children, we prefer using their involvement and take housework as dummy: 1 = yes.
[5] Dummy: 0 = illness problem, 1 = good health.
[6] Dummy: SP=0, Agro-pastoral area = 1.

factor is also women's time spent on livestock which is negatively correlated with their earnings from crops ($P<0.01$). Wives' education has a positive correlation with their earnings while with a higher level of males' education, women earn less from crops, probably through decreased males' participation in farming and thus less earnings from crops. Also, better-educated wives are more likely to be involved in other, such as off-farm, activities. Table 3.14 shows that with higher education, women earn more from off-farm activities. The significant effect of schooling on crop activities is in line with the work of Quisumbing (2007) that argued that return to schooling for both men and women are significant in agriculture, and contrast with other studies (Ilahi, 2000) which underline that better-educated women are less involved in farming. However, even if wives' education leads to higher individual earnings from crops, results regarding the household income show that a better-educated wife is less probably interested in growing crops (AP area). Moreover, the male's education significantly decreases ($P<0.05$) the female's participation in livestock activities. Other results from Pakistan (Kandker, 1988) found husbands' education among the significant determinants of women's time allocation (in off-farm employment) in Bangladesh. The distance from the market is negatively associated with earnings from crops ($P<0.001$): one% increase in distance leads to 0.70% decrease in earnings. This result may be explained by the fact that cash crops are mostly sold in the big cities (Dakar, Kaolack) that are remote from most villages and transportation costs cause lower earnings and benefits. Regarding housework,

Table 3.14. Comparison between off-farm earnings (CFA)[1] and level of wives' education in the year of the survey.

Level of Education	Off-farm earnings (CFA)	
	Mean	Std
Illiterate and basic level	41,377	10,030
Primary school level	54,166	44,944
Secondary school level	103,750	98,804

[1] 1 US$ = 475 CFA during the year of the survey.

the time that girls spent cleaning, fetching wood and water, and caring, positively predict wife's earnings ($P < 0.05$).

Regarding livestock (Table 3.15), women's time allocation predicts their income from livestock quite well ($P < 0.001$). Women's time use in feeding, milking, and water collection, and boys' marketing milk are good predictors of women's income from livestock, highlighting that women's time allocated to livestock tasks are productive factors of livestock income. In fact, for females this time includes almost all activities related to livestock. Evidence from peasant households in the Peruvian Sierra suggests a gender division of labour with females spending more time on livestock than men do (Jacoby, 1991). Regarding household tasks, negative effects on livestock outcome are found for women's involvement in chores ($P < 0.01$).

Important determinant factors in the probability of wives' participation in livestock activities are females' savings and cash transfers from migrants (positive effect). These increase the probability of their participation in livestock activities but the cash transfers decrease their participation in keeping livestock) (both at $P < 0.01$). These results reinforce our previous findings in the SP area where cash transfers have a negative effect on livestock activities, suggesting that such alternative earnings push women to involvement in other activities. Wife's savings also have a positive effect on females' income from livestock ($P < 0.05$). For the wife's savings, the positive effect (in both the outcome and the selection equations) can be explained by the fact that savings can mitigate risks and reinforce the investment of income from keeping livestock, which is an important form of savings for women in West African countries. For example, domestic dairy production is an activity mainly managed by women. They are responsible for marketing and the traditional processing of milk that results in yoghurt and butter. Wife's savings are thus an important determinant of livestock assets and milk production. Research has shown that livestock production, especially smallholder dairy farming, contributes to income generation of smallholder

Table 3.15. Heckman's two-step selection model; dependent variable: wife's income from livestock.

	Coefficient	Std	t-value
Outcome equation			
Husband's education [1]	-0.1239	0.0962	-1.29
Wife's education [1]	0.0559	0.1220	0.46
Wife's access to land [2]	0.5421	0.2634	2.06*
Household size (log)	0.3877	0.2265	1.71
Number of goat (log)	-0.0730	0.0806	-0.91
Number of sheep (log)	0.0127	0.0785	0.16
Number of cattle (log)	0.0894	0.0981	0.91
Time in livestock in log (W)	0.7435	0.1304	5.7***
Time spent in crops (W)	-0.1153	0.0809	-1.43
Zone [6]	-0.8285	0.4259	-1.95*
Wife's access to credit [3]	-0.1798	0.2084	-0.86
Land cultivated	0.0507	0.0578	0.88
Time marketing animals (H)	0.0142	0.1620	0.09
Time marketing animals (B)	0.3734	0.1505	2.48**
Time in livestock in log (H)	-0.0497	0.0764	-0.65
Time in livestock in log (B)	-0.1397	0.0875	-1.60
Time in livestock in log (G)	-0.0196	0.1108	-0.18
Time in livestock in log (Helpers)	-0.1016	0.0907	-1.12
Housework (H) [4]	-0.0292	0.0699	-0.42
Housework (W) [4]	-0.2796	0.0982	-2.85**
Housework (G) [4]	-0.0536	0.0539	-0.99
Cash transfers	0.0463	0.0506	0.92
Off-farm earnings	-0.0277	0.0207	-1.34
Wife's savings (log)	0.0541	0.0252	2.15*
Distance to market (log)	-0.2324	0.0763	-3.04**
Housework (B) [4]	0.0777	0.0912	0.85
Housework (O) [4]	0.0435	0.0879	0.49
Heath status (W) [5]	-0.5701	0.4477	-1.27
Constant	13.2988	1.7269	7.70
Selection equation			
Wolof group [7]	0.6061	0.2770	2.19*
Peulh [8]	0.7515	0.3259	2.31*
Husband's education	-0.1584	0.0810	-1.96*
Wife's education	-0.0619	0.1168	-0.53
Wife's access to land	0.3937	0.2431	1.62
Zone [6]	-0.0679	0.2707	-0.25
Cash transfers from migrants (log)	-0.0565	0.0195	-2.89**

Table 3.15. Continued.

	Coefficient	Std	t-value
Selection equation (continued)			
Off-farm wages	0.0320	0.0289	1.11
Wife access to credit	0.1877	0.2275	0.82
Wife's savings (log)	0.0537	0.0180	2.99**
Constant	-1.9340	1.2284	-1.57
Lambda	0.5907	0.6550	0.90
Rho	0.6576		
Sigma	0.8983		
N=250			
Censored observations=99			
Uncensored observations=151			
Wald $chi^2(31)=169.22$			
Prob $chi^2<0.0001$			

*$P<0.05$, **$P<0.01$, ***$P<0.001$.

H = husband, W = wife, B = boy, G = girl, O = other dependent members.

[1] Education dummy: 0= lowest, 1 = highest education.

[2] Dummy wife access to land: 1 = yes.

[3] Dummy wife's access to credit: 1= yes.

[4] Due to difficulties in reporting hours in house work by farmers and their children, we prefer using their involvement and take housework as dummy: 1 = yes.

[5] Dummy: 0 = illness problem, 1 = good health.

[6] Dummy: SP=0, Agro-pastoral area = 1.

[7] Dummy Wolof (Wolof=1, Sereer =0).

[8] Dummy Peulh (Peulh=1, Sereer=0).

farmers (Delgado *et al.*, 1999). It contributes to the capital accumulation of households that are poor in resources. According to Staal (2002), the income partially accrues to women and has a significant effect on the family's welfare (child welfare and nutrition). Ethnicity and area also are correlated with women's earnings. In the AP area, earnings from livestock are 0.82% lower than in the SP area.

Distance from the market also is a negative predictor of female earnings in livestock. One percent increase in distance leads to 0.23% decrease in women's income from livestock ($P<0.01$). In fact, distance from the market is important in marketing milk and this activity provides the largest share of female income (33% of the wife's income in our study). Female earnings highly depend on milk sold in the market; hence long distance to the market has a decreasing effect on income, more so because dairy products do not keep as long as cereals and cash crops (peanut, millet and water melon). Because of its important effect on income generation, employment and welfare for all categories of producers (especially the women and landless), governments in many developing countries (Kenya, India, China) have developed special policies to improve milk supply. These policies relate to technical, but also to institutional aspects (infrastructure) and organizational aspects, like the New Hope Dairy Group in China (World Bank, 2004). In fact, infrastructure is an important factor in the development process. In the case of Senegal, the lack of daily markets in rural areas (most villages have only weekly markets) cause a substantial loss of milk (products), especially during the rainy season. These aspects disfavour especially wife's earnings in livestock.

Determinants of husbands' earnings from farming

Off-farm earnings of women have a positive effect on husband's participation in growing crops, thus strengthening the findings that other complementary income allows farmers to secure their crop activities (Table 3.16): every additional franc from off-farm activities (on the logarithmic scale) increases the probability of the husband's participation in crop activities ($P<0.01$). Household size also is a good predictor of such probability. Distance to the market significantly decreases income: 1% increase in distance leads to 0.31% decrease of income ($P<0.001$). This effect reflects the role of infrastructure in agricultural productivity. In fact, earnings from crops depend highly on timely marketing. Since the collapse of West-African cooperatives in the 1980s after structural adjustment, most farmers resorted to informal channels for marketing their crops. In the research area, the markets of Kaolack in the AP area and Touba and Dahra in the SP area, are very dynamic in the improved marketing of cash crops and food crops. The proximity of markets also allows for better information about the prices of products. Moreover, transportation costs remain an important determinant of profits. Higher transportation costs reduce farmers' profits and cause losses due to transportation constraints and delays in the delivery of perishable products.

Time allocated by other helpers to household tasks is a negative factor in crop income because of the need for household labour to participate in crop activities (boys and other helpers). In the income equation, time spent by boys on crop activities is a negative factor because as we already noticed, boys are more productive in other activities (livestock). Also, as expected, off-farm activities are good predictors of the probability of growing crops. In the outcome equation, husbands' good health induces 0.80% more earnings ($P<0.05$): than in bad health.

Turning to men's participation in livestock (Table 3.17), positive factors are ethnicity (Peulh are more involved in livestock than other ethnic groups) and millet cultivation

Table 3.16. Heckman's two-step selection model; dependent variable: husband's income from crops (in log).

	Coefficient	Std	t-value
Outcome equation			
Age square (H)	-0.0007	0.0005	-1.28
Husband's age (log)	0.0743	0.0528	1.41
Husband's education [1]	-0.1101	0.0653	-1.69
Wife's education [1]	0.0319	0.1180	0.27
Wife's access to land [2]	-0.1774	0.1762	-1.01
Wife's access to credit [3]	-0.2734	0.1507	-1.81
Household size (log)	0.1480	0.2353	0.63
Water melon (yield in log)	0.0633	0.1355	0.47
Peanut (yield in log)	0.4632	0.1381	3.35**
Millet (yield in log)	-0.1812	0.1415	-1.28
Beans (yield in log)	-0.0002	0.2369	0.00
Number of goat/sheep log)	0.0118	0.0735	0.16
Number of cattle (log)	-0.0174	0.0705	-0.25
Time in crops in log (H)	-0.0028	0.0992	-0.03
time spent in crops in log (W)	0.0020	0.0687	0.03
Time in crop in log (B)	-0.4709	0.0986	-4.77***
Time in crop in log (G)	0.0452	0.0858	0.53
Time in crops in log (O)	0.0585	0.0614	0.95
distance to market (log)	-0.3147	0.0718	-4.38***
Housework (H) [4]	-0.0470	0.0736	-0.64
Housework (W) [4]	-0.1266	0.0734	-1.73
Housework (B) [4]	-0.1794	0.0900	-1.99*
Housework (G) [4]	0.0702	0.0469	1.50
Housework (O) [4]	-0.1153	0.0445	-2.59**
Health status [5]	0.8002	0.4000	2.23*
Constant	10.1395	1.5415	6.58
Selection equation			
Husband's education	-0.0149	0.1020	-0.15
Wife's education	-0.2771	0.1360	-2.04*
Off-farm earnings (log)	0.0470	0.0196	2.39**
Cash transfers from migrants (log)	0.0255	0.0361	0.71
Household size (log)	0.8932	0.3230	2.77**
Number of goat/sheep (log)	-0.2516	0.0936	-2.69**
Number of cattle (log)	0.0765	0.1123	0.68
Housework (H)	-0.0180	0.1052	-0.17
Housework (W)	0.1278	0.1083	1.18

Table 3.16. Continued.

	Coefficient	Std	t-value
Selection equation (continued)			
Housework (B)	0.3117	0.1388	2.25
Housework (G)	-0.0229	0.0761	-0.30
Housework (O)	0.5085	0.1974	2.58**
Health status	0.0680	0.5410	0.13
Constant	-2.2385	1.4281	-1.57
Rho	-0.9217		
Sigma	0.8020		
Lambda	-0.7392	0.2434	
N=249			
Censored observations==100			
Uncensored observations=149			
Wald $chi^2(31)=201$			
Prob $chi^2<0.0001$			

*$P<0.05$, **$P<0.01$, ***$P<0.001$.

H = husband, W = wife, B = boy, G = girl, O = other dependent members.

[1] Education dummy: 0= lowest, 1 = highest education.

[2] Dummy wife access to land: 1 = yes.

[3] Dummy wife's access to credit: 1= yes.

[4] Due to difficulties in reporting hours in house work by farmers and their children, we prefer using their involvement and take housework as dummy: 1 = yes.

[5] Dummy: 0 = illness problem, 1 = good health.

($P<0.01$). In fact, cultivated millet does not compete with livestock activities in terms of time allocation, like, for example, cash crops that need more care. Land cultivated with peanut is negatively correlated with the probability of participation in livestock activities. In addition, millet is used for consumption by both animals and humans. The AP area is negatively associated with participation in livestock activities. Because of the scarcity of available land in this area, keeping livestock remains a real problem. Cash transfers are also negatively associated with male participation in livestock. This strengthens our previous findings for the wife: cash transfers negatively contribute to keeping livestock ($P<0.01$). Cash transfers reduce the probability of participation in keeping livestock. In the livestock outcome equation, positive predictors for men are ethnicity and the number of cattle ($P<0.001$). Again, an important negative predictor is the distance to the market ($P<0.001$). This variable reduces the husband's income in livestock by 0.23% for each 1% greater

Table 3.17. Heckman's two-step selection model; dependent variable: husband's income from livestock.

	Coefficient	Std	t-value
Outcome equation			
Wolof [7]	0.5917	0.2332	2.54**
Peulh [8]	1.0042	0.2594	3.87***
Husband's education [1]	0.2014	0.0541	3.73***
Wife's education [1]	-0.0505	0.0803	-0.63
Household size (log)	0.0162	0.1784	0.09
Number of cattle (log)	0.3580	0.0701	5.1***
Number of goats (log)	0.0434	0.0614	0.71
Number of sheep (log)	0.1073	0.0605	1.77
Number of horses and donkeys (log)	0.0524	0.0794	0.66
Time marketing animals in log (H)	-0.0191	0.1291	-0.15
Time marketing animals in (B)	0.0860	0.1363	0.63
Time in livestock in log (H)	-0.0333	0.0565	-0.59
Time in livestock in log (W)	0.0289	0.0953	0.30
Time in livestock in log (B)	-0.0886	0.0635	-1.40
Time in livestock in log (G)	-0.1498	0.0951	-1.58
Time in livestock in log (Helpers)	0.0213	0.0729	0.29
Distance to market (log)	-0.2399	0.0615	-3.9***
Housework (H) [4]	-0.1175	0.0593	-1.98*
Housework (W) [4]	0.0220	0.0603	0.36
Housework (B) [4]	-0.0564	0.0736	-0.77
Housework (G) [4]	0.0733	0.0422	1.74
Housework (O) [4]	-0.0339	0.0508	-0.67
Off-farm earnings (log)	0.0149	0.0104	1.43
Cash transfers from migrants (log)	-0.0267	0.0171	-1.56
Health (H) [5]	-0.2898	0.2802	-1.03
Zone [6]	-0.4119	0.3026	-1.36
Constant	12.7590	1.2599	10.13
Selection equation			
Wolof [7]	0.1578	0.3513	0.45
Peulh [8]	0.9995	0.5085	1.97*
Husband's education	-0.1247	0.1171	-1.06
Wife's education	0.1054	0.1821	0.58
Wife's access to land	0.7569	0.4350	1.74
Peanut (yield in log)	-0.7923	0.3324	-2.38**
Millet (yield in log)	1.2410	0.3654	3.4**
Household size (log)	-0.3135	0.3930	-0.80
Zone	-1.5187	0.6264	-2.42**

Table 3.17. Continued.

	Coefficient	Std	t-value
Selection equation (continued)			
Off-farm earnings (log)	-0.0269	0.0276	-0.97
Cash transfers from migrants (log)	-0.0799	0.0257	-3.11**
Husband's health status	-0.0287	0.8119	-0.04
Constant	3.3979	2.2627	1.50
Lambda	0.2161	0.3373	0.64
Rho	0.2744		
Sigma	0.7876		
N =240			
Censored observations=34			
Uncensored observations=206			
Wald $chi^2(32)=205$			
Prob $chi^2<0.0001$			

$^{*}P<0.05$, $^{**}P<0.01$, $^{***}P<0.001$.

H = husband, W = wife, B = boy, G = girl, O = other dependent members.

[1] Education dummy: 0= lowest, 1 = highest education.

[2] Dummy wife access to land: 1 = yes.

[3] Dummy wife's access to credit: 1= yes.

[4] Due to difficulties in reporting hours in house work by farmers and their children, we prefer using their involvement and take housework as dummy: 1 = yes.

[5] Dummy: 0 = illness problem, 1 = good health.

[6] Dummy: SP=0, Agro-pastoral area = 1.

[7] Dummy Wolof (Wolof=1, Sereer =0).

[8] Dummy Peulh (Peulh=1, Sereer=0).

distance. This result strengthens the importance of market proximity in earning farm income. In the domain of livestock, men are especially involved in the marketing of animals. Likewise, the husband's involvement in housework (fetching wood and water) negatively affects his earnings in livestock. Regarding education, the results show that better-educated males earn more from livestock ($P<0.001$), thus strengthening our previous findings.

We also computed beta coefficients to test for economic significance of the variables; it appears that women's time allocated to livestock is an important factor of women's earnings since it is economically meaningful: it affects negatively earnings from crops (β = 0.87) and impacts positively on livestock earnings (β = 0.65). A change of one standard deviation of time allocated will result in 0.87 and 0.65 standard deviation change in women's earnings respectively for crop and livestock. Other significant determinants that are economically meaningful are wife's education, women's earnings from off-farm activities and household size: a change of one standard deviation of these variables lead to a 0.31, 0.33 and 0.28 standard deviation changes in women's earnings from crops, respectively. An economically significant effect is also noticed for wife's savings and distance from the market: a change of one standard deviation in savings and in distance from the market will result in 0.22 and 0.21 standard deviation change in women's earnings from livestock. Involvement of women and other dependent members in housework do not seem to be economically significant because the estimates of the beta coefficients are very small ($\beta = 0.09$ and 0.04 respectively, for women and for girls).

In concluding this section, estimations of determinant factors of individual incomes brought out differences across activities and gender. Moreover, women's and men's earnings from farming are especially associated with time allocated by females and boys to livestock activities, and that of girls and other dependent members to household tasks. These results confirm our expectations that women's time allocation is a productive factor in farming, and strengthens the finding that housework has an effect on female's earnings. Girls are the complement of their mothers in doing most household chores and this allows mothers to be more efficient in agricultural activities. Looking into the beta coefficients allows us to say that most of variables are economically significant in determining individual earnings; except for the involvement of women and other dependent members in housework, all variables show meaningful beta coefficients (larger than 0.20 in most cases).

Cash transfers from migrants have a negative correlation with the participation in farming both for husband and wife, while off-farm activities and wife's savings have positive correlations with husband's and wife's participation in farming. Distance to the market has a negative impact on all types of income procurement, both for husband and wife in their different activities. These results also suggest that male heads and wives secure farm activities, especially crops, through off-farm earnings and wives' savings. In fact, for rural households in Senegal, nowadays farming is a risky business because of decreasing rainfall and repetitive droughts. Consequently, the diversification of household activities through off-farm work allows for more participation in growing crops. Individual earnings also depend on area. Men gain more in growing crops in the AP area, where they participate less in livestock activities, while women earn less in growing crops there. While husbands' education has a significant effect on livestock earnings, better-educated wives earn more in crops. The latter results contrast with our previous finding that better-educated wives less probably participate in growing crops; but suggest that their better education may increase earnings. Also, participation of women in livestock activities seems to be influenced by males' education. Finally, health status is a good predictor of total earnings especially for men in crop production.

3.3.4 Individual time allocation determinants

Since labour supply of both husband and wife is an important input in all household activities (livestock, crops, and housework), time use in farming and its determinants were examined. We consider time used in farming by each of them in an activity x (T_{xi}), depends on a number of explanatory variables[6]: household size (H_z), housework involvement (T_h), productive inputs in the household like the number of livestock (L_{vi}) and land cultivated for each crop (L_c), assets of females or male heads in the household (G_{xi}) and other household characteristics Z (distance to the market, zone, credit access, wife's land access, wife's savings). Ethnicity and education have been used as dummies. For cultivated land, L_c, we consider that both husband and wife work in the same household plots (cash crops as well as food crops).

The input time use equations are:

$$T_{xi}= \alpha_0 H_z^{\alpha 1} T_{hi}^{\alpha 2} L_{vi}^{\alpha 3} L_{ci}^{\alpha 4} G_i^{\alpha_5} e^{\alpha_6 z} e^{u_3} \tag{3.10}$$

This can be converted into logarithms:

$$log(1+T_{xi})= \alpha_0 + \alpha_1 lnH_z + \alpha_2 ln(1+T_{hi}) + \alpha_3 log(1+L_i) + \alpha_4 log(0.1+L_c) + \alpha_5 log(1+G_i) + \alpha_6 Z + u_3 \,{}^{7} \tag{3.11}$$

For the time allocation Equation 3.10, Heckman's procedure is used for husbands in crop and livestock activities since 37.33% and 29.33% of observations are censored for the time allocated to crops and livestock activities, respectively. For females, the percentages are 48 and 43. The test for the joint null hypothesis that the coefficients on the variables common to both decisions are equal in the two equations, is rejected for both in the male heads' time in crop equation and in livestock equation (at $P<0.001$ and $P<0.01$, respectively). For women, the test rejects the null hypothesis in the crop equation at $P<0.10$.

[6] In our analysis we consider that there is a relationship between dependent and independent variables, although we neither assume nor test for causality of the relationship. Only for the sake of convenience we speak of independent variables 'affecting'or 'having an impact'on the dependent variable, where technically speaking we mean no more than 'association' or 'correlation'.

[7] We add 1 to the explanatory variables in order to avoid computational problems when taking the logarithm.

Women's time allocation's determinants

Table 3.18 shows that type of area is associated with the probability of women's time spent on crops ($P < 0.001$), which is 1.52% higher in the AP area than in the SP area. Wife's access to land also is a good predictor of her allocation of time to crop activities. As expected, food crop like millet favours a higher probability of women's participation in growing crops ($P < 0.01$). The area also matters because women are more active in crops in the AP area ($P < 0.001$). In the outcome equation, important factors are the involvement of other dependent variables in farming. Time use of girls and boys in crop growing are positively associated with time use of wives in crop growing (respectively at $P < 0.01$ and $P < 0.05$). For example, 1% increase in time use of girls in crops leads to 0.17% increase in female time used in crops. This highlights the important role of children in terms of labour force participation in the household. Also, women's caring for ill household members is negatively associated with their time in crop activities ($P < 0.05$), suggesting that illness of household members negatively affects women's time in farming. Women's time spent on crops is negatively correlated by the husbands' time in livestock suggesting that the wives work less in crops and increase their time allocated to livestock involving men. For crops, 1% increase in husbands' time spent, leads to 0.27% increase of wives' time. In fact, this result reflects the gender bias in women's labour force participation in crop production, because as we already showed (Chapter 2), wives earn less from cash crops than men (especially groundnut). Unless from the monetary compensation, wives earn 3% of total cash crop income and spend more time than husbands in activities related to groundnut cultivation (3.50 and 3.20 hours per day, respectively). This gender bias in wives' earnings per labour input is in line with other findings from developing countries (e.g. Smith *et al.* (1997) for Burkina Faso).

For livestock (Table 3.19), time use of girls and male heads is positively and significantly associated with female participation in farming (both at $P < 0.05$). That means that in households where male heads are involved in livestock, wives spend more time on it. This is also the case for crops. This result highlights the traditional patriarchal culture, where male household heads have power over women who must work in livestock production. Since most of the herders in our sample are Peulh, the results shed more light on their culture (see Chapter 2). This finding shows that livestock activities determine wives' time allocation to farming; Peulh are more involved in livestock ($P < 0.05$) than Wolof and Sereer. As for time allocation, in both areas women are constrained by their husbands in their participation in farm activities. As expected, a wife's savings predict her involvement in livestock because in most West-African countries, livestock constitutes the basic savings for females, and savings also enable her to expand the herd of sheep and cattle. The computation of the beta coefficients for testing economic significance of our variables shows that husband' s time allocated to crop and livestock activities are important factors in

Table 3.18. Heckman's two step selection model; dependent variable: wife's time input in crops (log).

	Coefficient	Std	t-value
Outcome equation			
Husband's education [1]	0.0030	0.0441	0.07
Wife's education [1]	0.0885	0.0735	1.20
Wife's access to land [2]	-0.2785	0.1627	-1.71
Household size (log)	0.2237	0.1534	1.46
Wife's acces to credit [3]	0.0104	0.1137	0.09
Number of goat/sheep (log)	0.0487	0.0598	0.81
Number of cattle (log)	-0.0206	0.0535	-0.39
Number of sheep (log)	0.0667	0.0516	1.29
Wife's caring ill members [4]	-0.3903	0.1719	-2.27*
Time in crop in log (H)	0.2710	0.0831	3.26**
Time in crop in log (B)	0.1335	0.0670	1.99*
Time in crop in log (G)	0.1701	0.0562	3.03**
Time in crop in log (O)	0.0266	0.0413	0.64
Time marketing animals (H)	0.0895	0.1032	0.87
Time marketing animals (B)	-0.1657	0.1131	-1.46
Time in livestock in log (H)	-0.1558	0.0703	-2.22**
Time in livestock in log (B)	-0.0374	0.0545	-0.69
Time in livestock in log (G)	-0.0677	0.0808	-0.84
Time in livestock in log (Helpers)	0.0397	0.0538	0.74
Housework (H) [4]	0.0121	0.0473	0.26
Housework (W) [4]	0.1289	0.0680	1.89*
Housework (B) [4]	0.0591	0.0627	0.94
Housework (G) [4]	-0.0386	0.0322	-1.20
Housework (O) [4]	-0.0347	0.0277	-1.25
Off-farm earnings (log)	-0.0123	0.0094	-1.30
Transfers from migrants (log)	0.0088	0.0117	0.75
Health status (W) [5]	-0.1222	0.3335	-0.37
Distance from markets (log)	0.0866	0.0584	1.48
Constant	1.8382	1.1377	1.62
Selection equation			
Wolof group [7]	1.1080	0.3284	3.37**
Peulh [8]	1.0312	0.3889	2.65**
Husband's education	0.0500	0.0959	0.52
Wife's education	-0.2312	0.1376	-1.68
Wife's access to land	0.6954	0.2756	2.52**
Peanut (yield in log)	0.0941	0.2069	0.45
Millet (yield in log)	0.5108	0.2151	2.37**

Table 3.18. Continued.

	Coefficient	Std	t-value
Selection equation (continued)			
Beans (yield in log)	0.5414	0.2909	1.86
Household size (log)	0.0640	0.2824	0.23
Zone [6]	1.5210	0.4094	3.72***
Off-farm earnings (log)	0.0195	0.0186	1.05
Ttransfers from migrants (log)	0.0368	0.0237	1.55
Constant	-5.3694	1.5660	-3.43
Lambda	-0.3404	0.2054	-1.66
Rho	-0.6688		
Sigma	0.5090		
N=246			
Censored observations=120			
Uncensored observations=126			
Wald $chi^2(32)=102$			
Prob $chi^2<0.0001$			

*$P<0.05$, **$P<0.01$, ***$P<0.001$.

H = husband, W = wife, B = boy, G = girl, O = other dependent members.

[1] Education dummy: 0= lowest, 1 = highest education.

[2] Dummy wife access to land: 1 = yes.

[3] Dummy wife's access to credit: 1= yes.

[4] Due to difficulties in reporting hours in house work by farmers and their children, we prefer using their involvement and take housework as dummy: 1 = yes.

[5] Dummy: 0 = illness problem, 1 = good health.

[6] Dummy: SP=0, Agro-pastoral area = 1.

[7] Dummy Wolof (Wolof=1, Sereer =0).

[8] Dummy Peulh (Peulh=1, Sereer=0).

explaining women' s time allocation to agricultural activities. One standard deviation change in hours spent by husbands in crop and livestock will lead respectively to 0.25 and 0.15 standard deviation change (increase) in women' s time allocated to crops and livestock.

Table 3.19. Heckman's two step selection model; dependent variable: wife's time input in livestock (log).

	Coefficient	Std	t-value
Outcome equation			
Husband's education [2]	0.0070	0.0439	0.16
Wife's education [2]	-0.0547	0.0673	-0.81
Wife's access to land [2]	-0.0552	0.1187	-0.46
Household size (log)	-0.0727	0.1492	-0.49
Number of sheep (in log)	-0.0084	0.0405	-0.21
Number of cattle (in log)	0.0854	0.0555	1.54
Number of goat (in log)	-0.0096	0.0497	-0.19
Time in crop in log (husband)	-0.0343	0.0609	-0.56
Time in crop in log (B)	-0.0390	0.0681	-0.57
Time in crop in log (G)	-0.0845	0.0721	-1.17
Time in crop in log (helpers)	-0.0382	0.0470	-0.81
Time in livestock in log (H)	0.1265	0.0660	1.92*
Time in livestock in log (B)	-0.0041	0.0537	-0.08
Time in livestock in log (G)	0.1385	0.0624	2.22*
Time in livestock (O)	-0.0688	0.0623	-1.10
Land cultivated (in log)	0.0101	0.0329	0.31
Time in crop (W)	0.1029	0.0564	1.83
Housework (H) [4]	-0.0237	0.0396	-0.6
Housework (W) [4]	0.0355	0.0538	0.66
Housework (B) [4]	0.0002	0.0517	0.00
Housework (G) [4]	0.0118	0.0315	0.37
Housework (O) [4]	0.0131	0.0466	0.28
Zone [6]	0.1025	0.2556	0.40
Health status (W) [5]	-0.3501	0.2591	-1.35
Outcome equation			
Distance from market (log)	0.0072	0.0449	0.16
Constant	2.1916	0.7805	2.81
Selection equation			
Wolof group [7]	0.7480	0.3432	2.18*
Peulh group [8]	1.0371	0.3911	2.65**
Education (H)	-0.1093	0.0890	-1.23
Education (W)	-0.1285	0.1363	-0.94
Peanut (yield in log)	-0.4058	0.2130	-1.90
Millet (yield in log)	0.3409	0.2524	1.35
Bean (yield in log)	-0.6260	0.3514	-1.78
Zone [6]	-0.6865	0.3779	-1.82

Table 3.19. Continued.

	Coefficient	Std	t-value
Selection equation (continued)			
Off-farm earnings (in log)	-0.0270	0.0188	-1.43
Wife's savings (log)	0.0736	0.0207	3.56***
Wife's access credit [3]	-0.4149	0.2627	-1.58
Cash transfers (log)	-0.0394	0.0217	-1.82
Health status (w) [5]	0.7407	0.5050	1.47
Constant	-0.6634	1.3835	-0.48
Lambda	-0.2502877	0.1869	-1.34
Rho	-0.52007		
Sigma	0.48125338		
N=235			
Censored observations=103			
Uncensored observations=132			
Wald $chi^2(28)=46.67$			
Prob $chi^2<0.0001$			

*$P<0.05$, **$P<0.01$, ***$P<0.001$.

H = husband, W = wife, B = boy, G = girl, O = other dependent members.

[1] Education dummy: 0= lowest, 1 = highest education.

[2] Dummy wife access to land: 1 = yes.

[3] Dummy wife's access to credit: 1= yes.

[4] Due to difficulties in reporting hours in house work by farmers and their children, we prefer using their involvement and take housework as dummy: 1 = yes.

[5] Dummy: 0 = illness problem, 1 = good health.

[6] Dummy: SP=0, Agro-pastoral area = 1.

[7] Dummy Wolof (Wolof=1, Sereer =0).

[8] Dummy Peulh (Peulh=1, Sereer=0).

Husband's time allocation determinants

Table 3.20 contains the results of the estimation of Equation 3.11 for men's time spent on crop activities. As we expected, time spent by boys in crops is negatively associated with husbands' time allocated to crops ($P<0.001$), because are complement of the male heads in crop activities. This means that boys are good helpers of the husband in terms of labour force participation. One% increase in boys' time in crops leads to a 0.47% decrease in husbands' time allocation in crops. Off-farm opportunities is positively correlated with the husband's time spent in crops ($P<0.01$). In fact as we

Table 3.20. Heckman's two step selection model; dependent variable: husband's time input in crops (log).

	Coefficient	Std	t-value
Outcome equation			
Age square (H)	-0.0007	0.0005	-1.28
Husband's age (log)	0.0743	0.0528	1.41
Husband's education [1]	-0.1101	0.0653	-1.69
Wife's education [1]	0.0319	0.1180	0.27
Wife's access to land [2]	-0.1774	0.1762	-1.01
Wife's access to credit [3]	-0.2734	0.1507	-1.81
Household size (log)	0.1480	0.2353	0.63
Water melon (yield in log)	0.0633	0.1355	0.47
Peanut (yield in log)	0.4632	0.1381	3.35**
Millet (yield in log)	-0.1812	0.1415	-1.28
Beans (yield in log)	-0.0002	0.2369	0.00
Number of goat (log)	0.0118	0.0735	0.16
Number of sheep (log)	-0.0434	0.0676	-0.64
Number of cattle (log)	-0.0174	0.0705	-0.25
Time in crop (H)	-0.0028	0.0992	-0.03
time spent in crops (W)	0.0020	0.0687	0.03
Time in crop in log (B)	-0.4709	0.0986	-4.77***
Time in crop in log (G)	0.0452	0.0858	0.53
Time in crop in log (O)	0.0585	0.0614	0.95
Distance to market (log)	-0.3147	0.0718	-4.38***
Housework (H) [4]	-0.0470	0.0736	-0.64
Housework (W) [4]	-0.1266	0.0734	-1.73
Housework (B) [4]	-0.1794	0.0900	-1.99**
Housework (G) [4]	0.0702	0.0469	1.50
Housework (O) [4]	-0.1153	0.0445	-2.59**
Health status (H) [5]	0.8002	0.4000	2.30*
Constant	10.1395	1.5415	6.58
Selection equation			
Husband's education	-0.0149	0.1020	-0.15
Wife's education	-0.2771	0.1360	-2.04*
Off-farm earnings (log)	0.0470	0.0196	2.39**
Cash transfers from migrants (log)	0.0255	0.0361	0.71
Household size (log)	0.8932	0.3230	2.77**
Number of goat (log)	-0.0127	0.1137	-0.11
Number of sheep (log)	-0.2516	0.0936	-2.69**
Number of cattle (log)	0.0765	0.1123	0.68

Table 3.20. Continued.

	Coefficient	Std	t-value
Selection equation			
Housework (H)	-0.0180	0.1052	-0.17
Housework (W)	0.1278	0.1083	1.18
Housework (B)	0.3117	0.1388	2.25*
Housework (G)	-0.0229	0.0761	-0.30
Housework (O)	0.5085	0.1974	2.58***
Health status (H)	0.0680	0.5410	0.13
Constant	-2.2385	1.4281	-3.57
Rho	-0.9217		
Sigma	0.8020		
Lambda	-0.7392	0.2434	
N=225			
Censored observations=85			
Uncensored observations=140			
Wald $chi^2(29)=130$			
Prob $chi^2<0.0001$			

*$P<0.05$, **$P<0.01$, ***$P<0.001$.
H = husband, W = wife, B = boy, G = girl, O = other dependent members.
[1] Education dummy: 0= lowest, 1 = highest education.
[2] Dummy wife access to land: 1 = yes.
[3] Dummy wife's access to credit: 1= yes.
[4] Due to difficulties in reporting hours in house work by farmers and their children, we prefer using their involvement and take housework as dummy: 1 = yes.
[5] Dummy: 0 = illness problem, 1 = good health.

have already observed, off-farm earnings permit reinvestment in crops (seeds and machinery) and thus increase the probability of husband's time allocated to crop activities (concisely put, off-farm earnings increase the probability of husband's time allocated to crops ($P<0.01$). The same positive effect is noticed for household size. As expected, the involvement of boys and girls in housework significantly decreases the husband's time spent in crops ($P<0.05$ and $P<0.01$, respectively). The negative effect of distance from market already observed in the previous sections, testifies to the time lost in farming because of long distance from markets and big centres. As expected, good health increases time allocation to farming ($P<0.05$) and renders husbands more productive. However, better- educated wives seem to push their husbands out of farming (negative effect at $P<0.05$).

For time allocated to livestock (Table 3.21), a positive factor is the time allocated to marketing ruminants (both for husbands and boys). For example, a 1% increase in time spent on marketing livestock, leads to high variation (0.98% in terms of elasticity) of husbands' time spent on livestock. This means that husbands' time spent on marketing ruminants constitutes an important labour input in livestock ($P<0.001$). However, the time spent by other dependent members (boys) on livestock activities reduces the husband's involvement in such activities ($P<0.05$). And in contrast, boys' time allocated to crops increases the husband's time allocated to livestock ($P<0.01$), suggesting that boys are productive in livestock and crops. Wife's involvement in housework increases husband's time allocation in livestock. In the selection equation – as expected – cash transfers reduce the husband's participation in livestock. Because women are a positive factor in the labour supply to livestock, their access to land (for crop growing) increases the probability that husbands spend more time in livestock ($P<0.05$). While education has a positive association with husbands' earnings as we have seen in previous sections, it has a negative correlation with their time allocated to livestock ($P<0.01$), probably because of the use of new technologies and their involvement in other activities (off-farm jobs). One percent increase in husband's schooling leads to a 0.17% decrease of his time in livestock. Likewise, better-educated wives lead to more time spent by husbands in livestock ($P<0.05$). In fact, better-educated wives are less interested in farming and that increases the males' burden in livestock activities. For husbands economically meaningful effects are noticed for the time that boys spent in crops and that husbands spent in marketing ruminants. For example, a change of one standard deviation in hours of these variables will result respectively in 0.48 and 0.66 standard deviation changes of husband time allocated to crops. Also, distance from the market has a relatively high negative effect on husbands' time allocation to crop activities.

We conclude this section with a summary: time allocation differs across activities and is gendered. For crops, important factors are related to the type of land and the involvement of other dependent members in farming and housework (boys, girls). Growing crops implies more labour time input, especially for wives. The involvement of boys in farming is an important determinant of men's time allocation, because it affects the husband's time involvement in both crops and livestock and has economic significance. Also, for livestock activities, predictors of change in labour time input of either wife or husband, are time spent on marketing ruminants, the number of animals in the household, cash transfers available, and education. Also, wives' labour time input is significantly and positively affected by husbands' time spent on livestock, suggesting a sort of gender specialization in farming activities in terms of labour force participation. The effect is economically meaningful because it is substantial. While good health of male heads positively predicts time spent in crops, distance from the market has a negative impact on time allocated to crop activities. This latter finding strengthens our previous findings that distance negatively affects earnings. Also, better-educated male heads impact negatively on their time in livestock activities.

Table 3.21. Heckman 's two step selection model; dependent variable: husband's time input in livestock (log).

	Coefficient	Std	t-value
Outcome equation			
Husband's education [1]	-0.1729	0.0714	-2.42**
Wife's education [1]	0.1978	0.1047	1.89*
Wife's access to land [2]	0.2460	0.2373	1.04
Household size (log)	-0.0243	0.2515	-0.10
Wife's access to credit [3]	-0.2182	0.1949	-1.12
Number of cattle (log)	0.0465	0.0902	0.52
Number of sheep/goat (log)	-0.0293	0.0801	-0.37
Land cultivated (log)	-0.0920	0.0546	-1.69
Time spent in crops in log (W)	-0.0018	0.0813	-0.02
Time in livestock in log (Helpers)	-0.0526	0.0966	-0.54
Time in crop in log (B)	0.1959	0.0890	2.20*
Time in crop in log (G)	0.0718	0.1025	0.70
Time in livestock in log (B)	-0.1985	0.0893	-2.22*
Time in livestock in log (G)	0.0142	0.1137	0.12
Time marketing animals (H)	0.9813	0.1674	5.86***
Time marketing animals (B)	0.2451	0.1520	1.61
Time in livestock in log (W)	-0.0193	0.1173	-0.16
Housework (H) [4]	-0.0105	0.0754	-0.14
Housework (W) [4]	0.1739	0.0765	2.27*
Housework (B) [4]	-0.0023	0.0893	-0.03
Housework (G) [4]	-0.0560	0.0520	-1.08
Housework (other helpers) [4]	-0.0991	0.0856	-1.16
Off-farm earnings (log)	-0.0056	0.0128	-0.44
Cash transfers from migrants (log)	-0.0086	0.0290	-0.30
Distance from market (in log)	-0.1409	0.0763	-1.85
Health status (H) [5]	0.3949	0.3423	1.15
Constant	2.2246	1.2857	1.73
Selection equation			
Wolof group [7]	0.4088	0.3009	1.36
Peulh [8]	0.8522	0.3887	2.19*
Husband's education	-0.0378	0.0940	-0.40
Wife's education	-0.0145	0.1278	-0.11
Wife's access to land	0.6372	0.3064	2.08*
Peanut (yield in log)	0.0534	0.2299	0.23
Millet (yield in log)	0.0696	0.2666	0.26
Household size (log)	-0.3969	0.3210	-1.24

Table 3.21. Continued.

	Coefficient	Std	t-value
Selection equation (continued)			
Zone [6]	-1.4943	0.4279	-3.49***
Off-farm earnings (log)	0.0126	0.0206	0.61
Cash transfers from migrants (log)	-0.0740	0.0230	-3.22**
Constant	1.8226	1.6526	1.10
Lambda	0.3237	0.7243	0.45
Rho	0.3661		
Sigma	0.8840		
N=237			
Censored observations=70			
Uncensored observations=167			
Wald $chi^2(33)=117$			
Prob $chi^2<0.0001$			

*$P<0.05$, **$P<0.01$, ***$P<0.001$.

H = husband, W = wife, B = boy, G = girl, O = other dependent members.

[1] Education dummy: 0= lowest, 1 = highest education.

[2] Dummy wife access to land: 1 = yes.

[3] Dummy wife's access to credit: 1= yes.

[4] Due to difficulties in reporting hours in house work by farmers and their children, we prefer using their involvement and take housework as dummy: 1 = yes.

[5] Dummy: 0 = illness problem, 1 = good health.

[6] Dummy: SP = 0, Agro-pastoral area = 1.

[7] Dummy Wolof (Wolof = 1, Sereer = 0).

[8] Dummy Peulh (Peulh = 1, Sereer = 0).

3.4 Conclusion

In Chapter 3 we have investigated the structure of productivity (relationship between total farm earnings and important input factors) in rural Senegalese households. Men, women and children (especially young boys) provide important labour input in all farming activities in terms of time allocation and income procurement. Moreover, women and girls have a crucial role in housework. And we showed that young boys and male household heads are also involved in housework. The results suggest that income generation in farming is not driven by one household member, i.e. the male head, but that each household member separately earns income in the different areas considered. Despite their remarkable contribution in terms of labour and time allocation, women earn less in farming activities. For this reason, there is a sort of monetary compensation from husband to wife for her effort in labour supply. Even if male heads' earnings are dominant, both husbands and wives are market-oriented. In some activities, wives perform better, for example in milk production where their earnings attain 4.5% of total farming income compared to 1% for men. Moreover women are very dynamic in off-farm activities, which constitute 37% of income from non-agricultural activities. These activities earn 53% of off-farm income for male heads. For females, significant differences in earnings can be found between the two areas of research, suggesting differences in wives' labour market participation between the two areas.

Livestock production and crop production seem to be complementary and competitive in labour force participation and time use. For example, land cultivated for groundnuts reduces the participation in livestock activities, whereas millet has a positive effect on participation in livestock. Moreover, keeping more livestock, especially cattle, has a negative effect on participation in crop activities. For this reason, ethnicity is associated with the probability of participation in both activities (livestock and crops). In terms of labour supply, our results show that women and children are productive factors in crop and livestock production. For example, boys' marketing of ruminants predicts income from livestock, while for girls, time is more devoted to housework which positively affects household income. However, husband's caring for ill members negatively affects household earnings from livestock (in the AP area). For females, the effect of caring is positive on livestock income in the AP area; however, on individual income, women's involvement in housework has a negative effect (in both areas for livestock and crops).

It was also shown that education has an effect on farming systems. Men's education is positively associated with income from livestock; however, the effect is negative for crops (in the SP area). For women, better education leads to less participation in growing crops, but may increase individual earnings from crops. Nevertheless, better-educated wives are more involved in off-farm activities. Moreover, access to credit stimulates women who are more interested in other activities than farming (trade and small entrepreneurship). These activities enable them to earn more and to diversify revenues and mitigate risks from farming. The results show an active labour market for women outside the household (especially in cities, as domestic helpers, teachers, etc.), which could induce a reallocation of their labour supply to crops. In the past ten years, drought, climate and environmental change (scarcity of rainfall, prices of inputs, and lack of subsidies) rendered crop growing less efficient and led to more rural-urban migration. For most farmers, growing crops became very risky. Hence, alternative activities like off-farm employment offer safer opportunities for households. Also, in our research area's rural labour markets educated people are a minority (53% of men and 75% of women are illiterate). This could stimulate farmers to invest off-farm earnings in girls' education. Cash transfers from migrants are negatively associated with earnings from livestock. Indeed, cash transfers constitute an important alternative in mitigating production risks; pastoralists tend to be more involved in crop activities when cash transfers are available.

A comparison of the two areas points at a specialization of activities between the two areas. Time allocation in livestock and the number of cattle are important factors of production in the SP area, whereas land and labour supply are important ones in the AP area. Also, individual incomes show interdependency between crop and livestock activities, with productivity highly depending on boys' and wives' time inputs in livestock, savings, ethnicity and alternative earnings (cash transfers). Concerning labour time allocation, our strong finding is that women are not specialized exclusively in housework, but are involved in the labour market and constrained by the time they have to spend on housework, and by the time their husbands spend in livestock and crops. It seems that women are forced to supply more labour to farming if their husbands work more in livestock or in crops. An increase in the burden of housework negatively affects female time allocation to farming, hence forces them to be more productive in order to produce the same quantity. Also, the distance of most rural areas to peri-urban markets disfavours husbands' incomes from livestock and wives' income from crops, for both areas, SP and AP. With respect to the effect of health status on household production, results show that husbands' good health positively affects the household earnings from crops (AP area), and health problems negatively affect women's earnings from livestock (AP area).

It has also been demonstrated through the estimation of beta coefficients that most variables that are statistically significantly correlated with earnings from agricultural activities have also a significant economic meaning. For example women's time allocated to livestock and housework, men's time allocated to marketing ruminants and cash transfers from migrants have substantial meaning with respect to their association with earnings from agricultural activities.

Finally, this chapter has highlighted that productivity is gendered and farm and off-farm household earnings are affected by a number of exogenous and endogenous factors, differently for husbands and wives. These factors can be related to individual characteristics such as education, or to socio-economic factors like access to credit, savings and assets (land and livestock endowment), or to the structural environment (infrastructure). Overwhelmingly, the participation of women in livestock is determined by the time that men spend in livestock and crop activities, suggesting a sort of patriarchal culture in the two areas. However, this finding does not sufficiently support the claim that women in the SP area are constrained in earnings. Moreover, the estimation of earnings per zone shows that women in the SP area are wealthier.

Earnings finance household consumption and daily food provision. However, we would like to know how decisions are taken in the household, and who is responsible for expenditures and other decisions to be taken. Moreover, how can we measure women's' empowerment in the household and determine what is associated with their power? These are the issues of the next chapter.

Chapter 4
Household decision-making in rural senegalese farming

4.1 Introduction and theory

4.1.1 Introduction

The status of women in the household has been considered a critical issue in changing the social and economic position and the well-being of household members. The main argument is that women's greater influence and access to market work may enhance their earnings and thus increase household welfare and well-being. In fact, women's subordination to men (i.e. related to their role restricted to the domestic sphere) has been seen as the main cause of gender inequality and deficiency of policies implemented to alleviate poverty (Visvanathan *et al.*, 1997). Hence, more interest is now focused on the way in which women's empowerment may gain them more bargaining power and more responsibilities in the household through increased control over their income and resources (UNDP, 2008, Dollar and Gatti, 1999). In poor rural households, women's control over their income may induce more power in decision-making and consequently enhance consumption of household goods. This chapter analyzes the position of rural women in Senegalese households with respect to their role in the provision of food and health (care and treatment). Also, it attempts to determine the main proxies that can be related to their power.

4.1.2 Relevance and definition of important concepts

Theoretically, the Resource Bargaining School supports the argument that women's greater contribution in earnings leads to more equitable sharing of decision-making power (Rogers and Deboer, 2001) and more responsibilities in family work (Both *et al.*, 1984; Kelly and Shortall, 2002). The assumption behind this argument is that the partner who contributes more in monetary earnings, is entitled to greater bargaining power in decision-making. Consequently, women's empowerment may induce the expectation of social change at the level of the household. However, these arguments have been disputed by other scholars who consider that even when they are active in the labour market, women have little bargaining power to renegotiate their position in the household, for example, in terms of voice or childcare (Morris, 1990; Ciabattari, 2001). In the past decade, initiatives and advocacy from development agencies have been used to improve women's access to assets and income and thus to greater decision-making power in households. However, little impact has been noticed on their economic empowerment and their decision-making in the household. One serious problem was that power was only analyzed from an individual point of view (Kabeer, 1994) without any links to the household setting that may potentially impact decision-making processes. Moving from the individual decision-making focus to the household level improves understanding of the social and economic settings of gender relations (Wheelock *et al.*, 1996; Kabeer, 1994). For example, husband and wife may have different points of view on their participation in decisions concerning labour supply and the well-being of the household as a whole. Such decisions are embedded in husband's and wife's economic benefits, traditional or patriarchal norms, and community characteristics.

In Sub-Saharan countries, particularly in Senegal as shown in Chapter 3, husband and wife gain their income separately and each individual contributes to the household, attempting to attain food security and satisfying household needs. However, household decision-making implies interpersonal interest and institutional responsibility in order to attain the welfare and the well-being of the household. For instance, in the traditional organization of production and consumption in rural households, groups of persons belonging to extended families (one or several households) work together in the same fields and share the same residence unit, under the responsibility of the head of the family. At the household level, the household head has the social responsibility to fulfil the nutritional needs of all household members. The role of the head as food provider is embedded in different social rules depending on ethnicity (Benoit-Cattin and Faye, 1982). For example in the Wolof community the head of the household is named *borom ndjël* and in the Peulh community, he is the *diom galle,* both meaning 'the one responsible for the household.' In the past, the favourable environment (climate, land and family labour availability), has benefited the head of the household to ensure his social obligations in managing household agricultural production and meeting household needs (through the produce from common fields, mainly millet, and the purchase of other goods not produced on the farm). The sudden disengagement of the State and the combination of many unfavourable factors (decreasing rainfall and soil fertility, population pressure and removal of subsidies), have reduced land productivity, and led to a dysfunction of the household head's role (Kelly *et al.,* 1996; Gaye and Kelly, 1996; Suwadu *et al.,* 2007). This chapter describes the relative status or position of women in this context and the trends in responsibility for meeting household needs. Moreover, it highlights the factors that might be associated with women's power as influenced by her bargaining power.

In the literature, women's bargaining power reinforces their overall empowerment at household level regarding their participation in consumption decisions (Pollak, 2005). Women's empowerment can be understood as incorporating three important inter-related dimensions: resource allocation (access, human and social resources), decision-making, and achievement (Kabeer, 1999). While decision-making processes are related to women's bargaining power, the dimension of achievement relates to the well-being of the household. In developing countries women's bargaining power has been mainly focused on their earnings gained or assets acquired during the marriage or brought with them into marriage, since their participation in the labour market is very low in terms of wages earned. Some important proxies have been used to construct a measure of females' bargaining power in the context of developing countries (Doss, 1997). In fact, the 'bargaining power' a member can exert based on the share of resources she contributed to the household, can be indicated by several proxies that vary across countries and regions. There is no single indicator of bargaining power because of the socio-economic and cultural environment in which women are embedded, in which factors such as education, religion, laws, economic independence, and socio-cultural norms, play a role. Several measures of bargaining power are used in the literature, like female share of income (Haddad *et al.*, 1997), assets brought into marriage (Duncan, 1990; Duncan *et al.*, 1997; Quisumbing and Maluccio, 2000), inherited assets (Quisumbing, 1994), current assets (Doss, 1996), land ownership and labour ability or education (Agarwal, 1997). Also, an important point is that any measure of bargaining power depends on the relevance of the analysis in a particular cultural environment (Meherun, 2006). Since there is no universal indicator of bargaining power, researchers try to adapt power indicators to the countries or societies concerned. Important determinants of intrahousehold bargaining power in rural families would be traditional rights to communal resources, traditional social support systems, and supports from the government and NGOs, which impinge on social norms and social perceptions (Agarwal, 1997). For the reasons above, the concept of 'women's empowerment' is often not clearly defined. As stated by Batliwala (1993:48) cited by Kabeer (1999): *'I like the term empowerment because no one has defined it clearly yet, so it gives us a breaking space to work it out in action terms before we have to pin ourselves down to what it means.'* So, the term 'empowerment' refers to process as well as outcome; empowerment leads to more bargaining and women's power leads also to more bargaining power. Since, we are not able to measure women's empowerment as a process, our attempt to measure it is by linking it to women's decision-making in food and health provision and to other human and socio-economic factors, such as mobility, education, assets ownership, credit and savings which may influence their current power. In the following text, when refering to the status of women, we mean their actual power (Figure 4.1).

We assume that women in Senegal play an important role in household decision-making through their capability to intervene in major decisions such as those on expenditure and those related to health or food provision. For example, women's power in decision-making may be associated with the health status and well-being of the household, because of their control over resources (earnings, expenditures). In fact decision-making concerning health expenditures would be an important determinant of the health status of the household. Who is the key household decision maker when it comes to resource allocation, especially with respect to food provision, health care and access to health facilities? Women's intervention in decision-making may be attributed to their share of assets (cattle, small ruminants), savings and

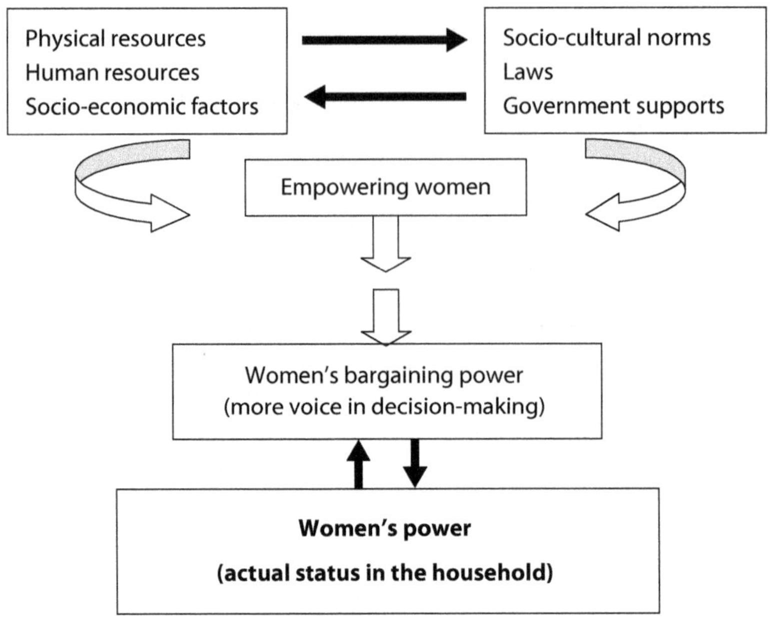

Figure 4.1. Illustration of the concept of empowerment and bargaining power.

unearned income (remittances), access to credit or micro-credit and contribution to expenditures. These different indicators of bargaining power may have an impact on her participation in household-decision-making. For this reason, difference in cultural behaviour and human-capital indices may affect women's power and their participation in the management of the household. Thus, we might expect differences between the Sylvo- and Agro-Pastoral areas in the actual power of women in terms of independence, voice and decision-making concerning consumption. Our hypothesis is that there is a positive relationship between the status of women in the household and their relative autonomy in terms of decision-making (concerning participation in food security, health and household well-being). In Sections 4.2.1 and 4.2.2, we describe the trends in decision-making and women's participation in expenditures on consumption and the roles played by husbands and wives in the provision of health care and daily food in the household. Furthermore, we use factor analysis to evaluate the determinants of the female power in the household. Section 4.2.3 deals with several empowerment proxies to determine important factors affecting females' decision-making in the household.

4.2 Decision-making in providing food and health care

This section provides insight into the responsibility for food and health provision both for husband and wife. For food, decisions are related to the responsibility in managing finance and providing food in different ways (borrowing, purchasing, exchanging assets for food, etc.). For health, the responsibility is related to the provision of care and the purchase of medicine.

4.2.1 Responsibility for food security and fulfilling household needs

Nowadays, in rural areas, many families have been fragmented due to migration of young people to the cities. Because of the shortage in labour supply, the head of the household is confronted with low productivity and poverty of his household and his role as provider of food is becoming more difficult. Moreover, trends in African rural households suggest that, in contrast with the past situation in which the husband was responsible for fulfilling household needs, currently both husband and wife try to ensure and provide daily food security in the household through agricultural and non-agricultural activities (Locoh, 1996; Pourcet, 1992; Pilon *et al.*, 1997; Fall *et al.*, 2004).

Results from our survey (Table 4.1) show that more than the wife, the husband still is responsible for providing consumption goods and for the management of household finance. Fifty-two percent of the households interviewed replied that the husband is exclusively responsible for the management of finance, 28% indicated that more than the wife, the husband is responsible, and 3% replied to the contrary, 10% reported that both husband and wife are equally responsible for finance management, and 6% replied that only the wife is responsible. Concerning the responsibility for expenditures, 51% report that the husband is the only food provider while 11% argue that both husband and wife are food providers. The exclusive responsibility of the wife for the provision of daily food and finance management is very rare: 0.7% and 6.3%, respectively. These results suggest that the husband as the head of the household is the major provider of food and the chief manager of finance. However, significant differences are noticed across zones for the management of finance, the responsibility for exchanging food, borrowing food for the household and exchanging household assets for food (Table 4.1)

The wife participates substantially in household decision-making, especially in the SP area where 15% of respondents declared that both husband and wife provide food equally, contrary to 8% in the AP area. Even in the management of finance, in the SP area wives are more involved than in the AP area, respectively 11.5% and 9.3% ($P < 0.001$). Table 4.1 shows that in all decisions related to the provision of daily food in the household even in difficult periods, wives have a voice and take part together with their husbands. Generally, households try to meet household needs through agricultural production and purchase of food at the market; however, 12% declared that the amount of food purchased and produced is frequently insufficient. In this case, many strategies are used for the survival of the household. Among these strategies, borrowing money or borrowing food from relatives constitutes an

Table 4.1. Contribution to food provision for the household in percentages by area.

	Management of finance		Who purchases food in the household		Borrowing money to buy food		Borrowing food for the family		Exchange household assets for food	
	SP area	AP area	SP area	AP area	SP area	AP area	SP area	AP area	SP area	AP area
Husband only	42.95	62.00***	39.60	62.91***	42.28	48.98	31.54	27.15**	2.01	6.62**
Husband more than wife	30.87	24.67	43.62	27.81	25.50	15.65	31.54	13.91	6.71	5.96
Both equally	11.41	9.33	14.77	7.95	7.38	6.12	10.74	13.91	1.34	4.64
Wife more than husband	3.36	2.67	1.34	0.66	2.68	0.68	0.67	2.65	1.34	3.97
Wife only	11.41	1.30***	0.67	0.66	2.68	1.36	3.36	0.66**	2.01	21.19
We never do that	-	-	-	-	19.46	27.21	22.15	41.72	86.58	57.62

T-test for difference between areas: ** $P<0.01$, ***$P<0.001$

important component (19% of the households). Both husband and wife are involved in borrowing money to buy food or borrowing food for the family: 7 and 12%, respectively. Strategies applied to cope with scarcity in food provision are mostly visible in the AP area with respect to borrowing food or exchanging household assets for food.

Other studies in Senegal showed that women are effectively involved and decide on several issues including food provision, energy provision (in particular the traditional source, i.e. wood), water collection and schooling (Ndiaye Diop, 2007). These results are in line with the fundamental role that women play in food provision, relating to their productive role (feeding the family and nurturing the work force) (Niehof, 2003). Decision- making according to this rural survey (Ndiaye Diop, 2007) depends on the financial situation of the household. In the poorest households, women tend to take most of the decisions regarding household needs because of their capability to diversify income through non-agricultural small-scale activities (small trade, entrepreneurship, etc.). In this situation, men mostly migrate to cities, looking for off-farm work, and women take most of the decisions in the household to make ends meet. Depending on who receives the cash from migrants, decisions in, and responsibility for food provision and expenditures are either taken by the wife or the husband's mother. In wealthier households, men take most of the decisions and women play a complementary role in the provision of consumption goods. In the latter case, women increase their capability for saving or accumulation of income, and develop other social-capital relationships.

4.2.2 Responsibility for dealing with health problems, and time allocated to health care

In many African societies, the inequity in resource allocation (mainly disfavouring women), decision-making regarding resource allocation, and human-capital endowment limits vulnerable people (women, children) in accessing health care and health facilities. For example, health problems lead to costs due to medical treatment. These costs can be seen as both direct (purchase of medical input at the market) and indirect through the loss of time (normally used in the labour market) spent by healthy household members to take care of the ill members. In our survey, decisions about health issues, regarding the type of medication, are mainly made by the husband (78% of cases), in just 2% of cases reported, health decisions are taken by the wife. However, according to the respondents, joint decisions on health care and type of medical treatment can also be made by husband and wife (20% of cases). These results are highlighted by those concerning the provision of cash paid for medical care or treatment of ill persons. The results show that in 78% of cases, the husband is the provider of the money and the wife only intervenes in 7% of cases. In some households (9% of cases) the money is provided jointly. Relatives from abroad or other persons intervene in 6% of the cases. However, when asking women whether they participate in health-cost decisions, 78% of them reported that they did. These results suggest that both make decisions about health issues but that the husband is mainly responsible for health-related costs, and women have limited resources to spend on health. These results are in line with the fact that when it comes to 'taking responsibility for health problems,' it is the man who dominates because of his access to resources (Niehof, 2004). Most of the time women are confined to their role as caregivers. Surveys undertaken by the Swiss Assistance Office in Senegal (Ndiaye Diop, 2007) reported that women mainly contribute to care (less costly) and less to more expensive health problems (injuries, broken leg, etc.) where men are the major decision makers.

When setting priorities regarding the destination of resources in the household, our interviews brought out that provision of food is always the first priority. According to the interviews, the decision to take medical treatment mostly depends on the financial situation of the household head and own willingness to face health problems. The wives can contribute voluntarily from their own income when it is available, but a priori husbands have to take care of their wives, children and parents in terms of health care and health costs. Apparently, health costs are not met through consensual decision-making or pooling the spouses' incomes, as in the case of food provision described in the previous section. Health costs are covered at the time of their occurrence by available money controlled by either husband or wife, but mostly by the husband. Decisions on health costs or medical treatment take into consideration alternative solutions not always requiring cash payments. For example, in case of illness, recourse to traditional healers is taken by 14% of households. Moreover, 16% decided not to apply for any treatment and keep their ill members at home. In fact, in some societies illness is considered a bad event nourished by undesirable spirits that can only be treated by God or healers. This reasoning also explains the fact that only 3% combine the modern treatment from the hospital with the traditional prescription from the healers. Respondents using traditional healers explain their choice from different causes, the most important one being that traditional healers are the cheapest (36% of respondents). Other arguments are the efficiency of the treatment (34%), inheritance of traditional beliefs from parents (19%), and miraculous results and the easy access to healers (10%). From these results, it appears that only 'serious health' problems merit the husband's attention. Pregnancies and reproductive diseases do not often merit medical attention from the husbands, and with such problems most women rely on traditional healers. The preference of alternative health care to modern treatment is also found in other studies of Sub-Saharan communities (Tipping *et al.*, 1995; Ngom *et al.*, 2000; Coppo *et al.*, 1992). For example, in Benin, traditional medicine is popular due the lower price and flexible arrangements (payment only if the patient is fully recovered) (Ngom *et al.*, 2000). According to 53% of the respondents, the reluctance to consult modern services (hospital and health centres) is due to the high cost of medicines. 25% always find the waiting time at the hospital very long, and the other respondents deplore the lack of personal treatment and equipment at the medical services.

When it comes to family planning and discussions on sexually transmitted diseases such as HIV/AIDS, 34% of partners' respondents reported that this issue is discussed by the couple and 35% exchange ideas and knowledge about HIV/AIDS. The other respondents (65%) do not discuss family planning and HIV/AIDS because only God decides about the number of children. Table 4.2 shows the lack of joint decision-making, of information exchange and discussions on the subject. These are mainly caused by: (1) insufficient knowledge about the disease, (2) norms and customs which do not permit discussing these issues, and (3) lack of interest in the subject.

These results suggest that knowledge about family planning in the household is very limited. Communication between husband and wife in this area is not favoured by local norms and customs, and women are mostly oriented to the use of methods of traditional medicine instead of modern ones.

A comparison between areas shows that more SP women (27%) made joint decisions with their husbands about health cost and health care, compared to 14% of AP women. The costs related to health have been handled by 12% of SP women compared to 5% of AP women. In the latter area, the husband is the main financial decision maker when it comes to health costs and health care (82% compared to 68% in the AP area). These results reinforce the evidence of the wife's power in the SP area where women are economically more comfortable.

As we observed in Chapter 3, the reallocation of women's time, for example, from the labour market to the care of ill members, may induce a loss of productivity and

Table 4.2. Self-reported causes for lack of discussion on family planning (percentage).

Causes for lack of discussions	%
Lack of knowledge	42.69
Norms and customs	17.55
Not interested in family planning	15.00
We are not concerned	13.00
We do not communicate	7.76
Family planning should not be allowed	4.00

income. The issue of ill health pertains not only to persons affected by a disease, but also to household members experiencing health problems related to pregnancy and child mortality, and chronic illness like malaria, flu or intestinal gastric diseases. Results of our survey show that in all cases of health problems reported, the average duration of illness is around 11 days for severe or acute diseases, and 52 days for chronic illness. In all cases of health problems reported, a disease induces a reallocation of time to the care of ill members. On average, the time for care-giving is around three hours per day. Care is mostly given by wives (44% of cases). Husbands gave care in 18% of the cases reported, followed by daughters (17%) and other relatives in the household (cousins, uncles or nephews: 11%). Boys provided care in 5% of the cases; their intervention is only due to the unavailability of wives or daughters. Elderly household members provide care for ill members only in a few cases (5%). Health care seems to be provided at the expense of women's daily household tasks and chores.

Daughters substitute or complement their mothers in daily occupations including care for ill members. These results reinforce the fundamental role of females in the 'private' domestic sphere of the household and the gender division of labour (Gardiner, 1997; Alston *et al.*, 1998; Whatmore, 1991; Niehof, 2004). Even in developed countries, women devote significantly more time performing health care than men, on average 14.2 hours per week compared to 7.5 hours for men (Zimmerman *et al.*, 1999). Particularly, the care for sick children with chronic illness is entirely within the sphere of women (Hill, 1994; Hill and Zimmerman, 1995). In fact, for women, health care is an obligation inspired by love and family responsibility. However, the time allocated to caring for ill members is of course time lost for the labour market (Shortall, 1992). This gendered division of labour where women are confined to household tasks and domestic occupations, reinforces the secondary status and the gender inequality prevailing in most rural communities (Whatmore, 1991; Moser, 1993). According to the literature (Momsen, 2004), the situation where females are almost entirely responsible for taking care of ill members is a consequence of their powerless position in the household, lack of voice, and limited participation in decision- making. Even for women who participate in the production system, their association with care-giving is explained in terms of their 'natural' maternal predispositions (Kabeer, 1994).

4.3 Measuring women power or relative status

This section describes the relative status of females by using factor analysis to detect important factors that might be related to their power. Further, using regression analysis, we shall try to identify the determinants of these factors[8].

4.3.1 Measurement of female relative status and its determinants

Women contribute to decision-making by using a kind of bargaining power that we try to capture through several variables like the wife's income, her access to credit, her savings and assets in the household, her off-farm wages and so on. However, as we saw in the previous section, it is well documented that the wife's bargaining power is multi-dimensional, and this makes it hard to measure. In this section we use self-reported information on women's relative status as an indicator of women's power. In doing so, we use different variables related to the social and economic situation of women in the household: some relate to sharing income, others to gender responsibility in food security (borrowing money or food) or the management of income (management of finance, buying food), while other variables are related to women's independence in making decisions on health issues (a woman's freedom to go to the hospital or the market without permission from the husband, freedom to decide on health costs, or freedom to discuss family planning with their partner, etc.). These variables capture different aspects of women's status, such as her mobility, her control over income and her independence in voice and in management. All information gathered about their status has been reported by women themselves, in the absence of their husbands, so we exclude a priori biased reporting of information due to the husband's influence.

Generally, the variables relating to sharing income could reflect the financial autonomy of the wife in terms of income. For example, in the context of developing countries, the wife's sharing of income with her husband is due to a sort of coercion or power that the husband exerts over his partner, consequently restricting her autonomy in decision-making. In contrast, the husband's sharing of income could indicate more or less independence for women in deciding on household expenditures (on clothes, health, food, etc.) and a degree of co-responsibility for expenditures in the household. The variables related to gender responsibility in food security include all variables used in Section 4.1 to describe the wife's participation in and control over expenditures that can explain the position of women in the household. The variables related to asking permission (to go to the market or to the hospital, or

[8] In our analysis we consider that there is a relationship between dependent and independent variables, although we neither assume nor test for causality of the relationship. Only for the sake of convenience we speak of independent variables 'affecting'or 'having an impact'on the dependent variable, where technically speaking we mean no more than 'association' or 'correlation'.

to buy medication) could reflect independence of women in their participation in the labour market either as seller at household level or as retailer, and reflect some independence in decision-making regarding important issues. In many West-African countries, women are highly involved in the market as sellers of vegetables, cereals, milk, fish or clothes (Boserup, 1970; Reardon *et al.*, 1992). This involvement leads to economic independence of women in the household and may influence their decision-making power within and outside the household.

In total 12 variables derived from our surveys have been used in a factor analysis to estimate n the household. Five of these variables are related to responsibility: (1) in managing finance, (2) in borrowing food, (3) in buying food, (4) in exchanging food, (5) in borrowing money to buy food. Two concern sharing: (6) husband shares income with his wife, (7) wife shares income with her husband. Four variables concern the wife's autonomy in taking action without permission: (8) to go to the market, (9) to take children to the hospital, (10) to go to the hospital for herself (11) to pay for medication, And the final autonomy variable is: (12) wife's autonomy in family planning decisions.

Factor analysis is used to discover unobserved factors (in our case power) explaining a larger number of variables or indicators. This technique intends to depict correlations between observed variables in order to estimate common factors. Furthermore, we investigate several variables that might explain women's power as indicated by the factors (reflecting the status of women in the household), in particular the husband's and the wife's incomes, her household savings, the value of her household assets, her access to credit, the household size and her involvement in off-farm activities. To control for endogeneity in the equation, we use other exogenous variables like religion in the Senegalese community, the polygamous status of the household, ethnicity, the type of area and husband's and wife's education. In fact, the prevailing Muslim religion in the context of Senegalese households implies a system of regulation of women's behaviour and status in the household, and ethnicity reflects cultural gender roles in decision-making affecting women. The area of research is associated with the livelihood system that the household is part of. Thus the estimation equation of the female status within the household (captured by different factors) can be written as depending on several variables as follows:

$$F_{ij}= \alpha_0+ \alpha_1 W_{vj}+ \alpha_2 H_{vj}+ \alpha_3 W_{ej}+ \alpha_4 H_{ij}+ \alpha_5 Z_j+ u_j \qquad (4.1)$$

where, F_{ij} denotes the i-th factor of women's relative status for individual j ($i=1,\ldots,n$, $j=1,\ldots,J$, n being the number of factors and J the number of individuals). W_{vj} is a vector of female characteristics like education, age, age squared, H_{vj} is a vector of husband characteristics (education, age, age squared), W_{ej} are the wife's bargaining power variables (share of income, savings, access to credit, physical assets and contribution to expenditures), H_{ij} is the husband's income, Z_j is a vector of household and community characteristics (like religion, ethnicity, area of living) and u_j is the error term.

To test the validity of our estimation, the closeness of fit for all specifications is indicated by an F-test for the joint significance of the variables. The Wald test is used to test the null-hypothesis that the coefficient of an explanatory variable equals zero.

4.3.2 Results of estimation

The participation of women in household decision-making may express their actual power. The self-reported information about their relative status reflects the wives' voice and independence in decision-making. For the responsibility for food provision and management of income, it has already been demonstrated that the husband is the major decision maker in all areas (see Section 4.2.1). Moreover, for the mobility variable, 81% of the women reported that they asked permission to go to the market. However, for the access to health treatment at the hospital and medication for women and children, mobility was less constrained. 42-50% reported not asking for permission to go the hospital, as follows: 50% went for their own health, 42% went to pay for medication, and 44% took their sick children to the hospital. It appeared that men and women shared the responsibility for health by contributing both money and care. This result is strengthened by the evidence suggesting that women as caregivers ensure a responsible role in their own health status and in the health of their children (caring and going to the hospital) (Caldwell, 1986; Basu, 1992; Meherun, 2006). A comparison between areas shows that AP women are more autonomous in mobility decisions because they ask less permission than

Table 4.3. Income-sharing and asking permission by gender and by area (percentage).

	SP area		AP area		Entire sample	
	Husband	Wife	Husband	Wife	Husband	Wife
Sharing income						
No sharing	2.68	19.59	1.32	14.29	2.00	16.95
Share small part	66.44	60.14	27.15	38.78	46.67	49.49
Share large part	29.53	19.59	64.24	30.61	47.00	25.08
Share all	1.34	0.68	7.28	16.33	4.33	8.47
Wife asking permission: yes						
To go to market		86.49		80.85		83.74
To go to hospital		58.11		43.84		5.021
To purchase medication		47.30		39.31		43.34
To take children to hospital		52.03		39.58		45.89

their SP counterparts. This result may be related to their important role in trading household products (cereals and vegetables) and the closer average proximity of AP households to the markets compared to SP households. This finding suggests that more than AP women, SP women are involved in decisions regarding expenditures as indicated in Section 4.2.1, but are more constrained in mobility than their AP counterparts (Table 4.3).

Regarding the income-sharing aspect, males were more willing to share their income than women; however, 46% of husbands and 50% of wives agreed to share only a 'small part' of their income. Only 17% of wives reported that they did not share their income, thus reflecting a sort of independence in the management of their income. This data shows that women in the SP area are more autonomous regarding monetary flows, because they share less of their income with their husbands than women in the AP area.

Taking into account the importance of these different women's status indicators in household decision-making, it will be relevant to study the determinants of these factors: are they related to their resources, their labour income (agricultural earnings) or non-labour income (cash transfers)? Are they related to their individual characteristics (savings, age or ethnicity) or other exogenous factors such as their

Table 4.4. Rotated component matrix (Varimax).

	Factor 1: permission	Factor 2: sharing income	Factor 3: responsibility food provision (in borrowing food and money)	Factor 4: management of income
Who manages finance in the household	-0.389	-0.546	0.134	0.477
Husband shares income with wife	-0.375	0.640	-0.377	0.318
Wife shares income with husband	-0.436	0.613	-0.433	0.136
Who usually buys food	-0.338	-0.295	-0.040	0.763
Who borrows food for the household	-0.230	0.443	0.660	0.160
Who borrows money	-0.031	0.382	0.766	0.096
Permission to pay for medicine at hospital	0.882	0.142	-0.064	0.257
Permission to go to hospital	0.827	0.097	-0.023	0.268
Permission to take children to hospital	0.872	0.118	-0.047	0.245

Areas shaded in grey indicate instances without permission.

education, their religion or their access to credit? This will be considered in the following sections.

The results of the factor analysis show the variances extracted by the factors indicated by the Eigen values (EV). Factor 1 accounts for 32% of the indicators' variance, factor 2 explains 18% of the variance, factor 3 explains 15% and factor 4 explains 13% (Table 4.4). Using Kaiser's Criterion considering only factors with eigenvalues greater than one, we retain four factors. Moreover, considering the graphical method of the Scree test proposed by Lattell (1966), we again retain four factors where the plotted eigenvalues show a kink after the fourth factor. After Varimax rotation of the Component Matrix, we can say that the first factor is highly related to the questions in connection with 'asking permission'. This first factor encompasses the wife's going to the market, her purchasing medication and her accompanying children to the hospital, in all instances without permission (shaded in Table 4.4). This factor can be called 'permission' and indicates the relative independence of women in this respect. The second factor captures opinions about the 'income-sharing' variables, indicating females' economic power. The third factor relates to responsibility in food provision (borrowing food and money). Finally, the fourth factor captures the management of finance in the household as an indicator of women's responsibility in this area. The factor analysis of the 12 variables resulted in four factors apparently describing the decision-making status of women in the sample. The status of women in terms of independence, mobility and responsibility in the household can be used to estimate the determinants of women's actual power in the household, including their income and savings, their assets and relative contribution to expenditures.

Table 4.5 shows the ordinary least square (OLS) estimates of the major determinants of women's relative status indicated by the four factors in the columns (permission, sharing income, responsibility for food provision and management of income in the household). It is expected that human and cultural as well as socio-economic factors will be associated with the four proxies: 'permission factor', 'husband sharing income', 'responsibility in borrowing food or money' and 'management of income'.

Important determinants of the permission factor are wife's savings ($P < 0.01$), wife's access to credit ($P < 0.05$), zone ($P < 0.05$), wife's expenditures on food and health ($P < 0.05$) and ethnicity ($P < 0.001$). Wife's savings and wife's access to credit had a decreasing effect on the permission factor; this means that a wife's savings allow for more power in decision-making, implying less constraint to go to the market, to the hospital or to purchase medication. These results suggest that a wife's savings and her access to credit in various forms [tontines (annuities shared by subscribers to a loan), valuable jewelry, micro-credit] induce more independence and autonomy concerning her mobility to increase her involvement in the labour market, or to take important decisions (e.g. to purchase medication for children or to go to the

Table 4.5. Results of OLS estimations of the four decision-making status factors.

Variables in log	Factor 1: permission	
	Coefficient	t-value
Age of husband	-0.0021	-0.24
Age of wife	0.0081	0.89
Transfers husband	0.0092	0.67
Male's off-farm income	-0.009	-0.40
husband's annual income	-0.037	-1.08
wife's annual income	0.0062	0.48
Wife's off-farm wages	-0.0015	-0.56
Transfers to wife	0.0497	1.95*
Wife's ownership of cattle	-0.0836	-1.16
Wife's sheep/goat ownership	0.1587	2.63**
Household size	-0.1967	-1.22
Wife's access to land [1]	-0.0057	-0.04
Wife'savings	-0.0301	-2.55**
Wife's access to credit [2]	-0.2765	-1.95*
Level of education husband [6]	0.0626	1.17
Level of education wife [6]	0.0058	0.07
Wife's expenditures on food	-0.0345	-2.31*
Wife's expenditures on clothes	0.0177	0.92
Wife's expenditures on school	0.0084	0.52
Wife's expenditures on health	-0.0282	-2.05*
Religion [3]	-0.0684	-0.16
Polygamy [4]	-0.0178	-0.41
Zone [5]	0.5099	2.34*
Wolof [7]	1.314	6.47***
Peulh [8]	1.146	5.00***
Constant	-0.5166	-0.46
Adjusted R^2	0.28	
Wald test for the rejection of the null hypothesis F-values	8.15***	

* $P<0.05$, ** $P<0.01$, *** $P<0.001$.
[1] Dummy (1 = yes).
[2] Dummy (1 = yes).
[3] Dummy (0 = Catholic, 1 = Muslim).
[4] Dummy (1 = yes).
[5] Dummy (0 = SP area, 1 = agro-pastoral area).

Factor 2: sharing income		Factor 3: responsibility food provision (borrowing food and money)		Factor 4: income management	
Coefficient	t-value	Coefficient	t-value	Coefficient	t-value
-0.0011	-0.12	-0.0228	-2.28	0.0134	1.47
0.0180	1.99*	0.0328	3.27**	-0.0185	-2.02*
-0.0106	-0.77	0.0150	0.99	-0.0232	-1.66
-0.0484	-2.08*	-0.0675	-2.62**	0.0109	0.46
0.0660	1.94*	-0.0495	-1.31	-0.0745	-2.15*
0.0015	0.12	-0.0261	-1.88	0.0247	1.94*
0.0063	0.39	0.0046	0.26	-0.0054	-0.33
-0.0595	-2.35*	0.0326	1.16	-0.0130	-0.51
0.0113	0.16	0.1075	1.36	0.0778	1.08
0.1494	2.51**	0.0076	0.12	-0.0316	-0.52
-0.0699	-0.44	-0.336	-1.90	-0.4141	-2.55**
0.2767	1.87	-0.3835	-2.33**	-0.2395	-1.59
0.0076	0.64	-0.0118	-0.89	0.0096	0.79
0.0423	0.30	-0.3845	-2.46**	-0.090	-0.63
0.0440	0.84	-0.0286	-0.49	-0.0623	-1.17
-0.0492	-0.61	0.0581	0.64	0.0021	0.03
-0.0192	-1.34	-0.0081	-0.50	0.0368	2.45**
-0.0145	-0.78	-0.0377	-1.79	-0.0149	-0.77
0.0415	2.59**	0.0041	0.23	-0.0059	-0.37
-0.015	-1.10	0.0018	0.12	0.0573	4.05***
0.6046	1.45	0.2338	0.51	-1.092	-2.57**
-0.0186	-0.36	-0.0583	-1.23	0.0155	0.36
-0.2831	-1.31	0.1461	0.61	-0.2443	-1.11
-0.5513	-2.74**	-0.4225	-1.90*	0.0670	0.33
-0.3862	-1.70	-0.2880	-1.14	0.0874	0.38
-2.235	-1.99*	2.5168	2.02*	4.461	3.90***
0.32		0.17		0.29	
6.49***		6.93***		9.20***	

[6] Dummy (0 = lowest, 1 = highest).

[7] Dummy Wolof (Wolof = 1, Sereer = 0).

[8] Dummy Peulh (Peulh = 1, Sereer = 0).

hospital). Findings in the literature show that 'permission' indexes have a positive and significant impact on children's health (Varadharajan, 2003).

Wives' larger contribution in expenditures on food reduces the need for asking permission, and this too, suggests a sort of power in decision-making. It is obvious that a female's contribution to food provision, which can be considered a household good, is a good argument for her ability to participate in the market. Ethnicity also helps to explain the factor 'asking permission'; Peulh and Wolof women more than Sereer women, have to ask for permission.suggesting that all women face mobility constraints. However, the factor 'asking permission' is more important in the AP area (where are living the Wolof group) than in the SP area. The R^2 of the regression equals 0.28 and the test of the null hypothesis that the coefficients all equal zero, is rejected at ($P < 0.001$). Looking into the standardized coefficients tell us that the wife's food expenditures on the 'permission factor' is economically meaningful because the estimate is 0.25. Also ethnicity show high estimates of the standardized coefficients (0.57 for Wolof and 0.49 for Peulh). This finding suggests that the correlation between ethnicity and the 'permission factor' has significant effects economically speaking.

For the factor 'sharing income', significant effects are found for the wife's age ($P < 0.05$), her ownership of sheep and goats ($P < 0.01$), husband's income ($P < 0.05$), remittances to the wife ($P < 0.05$), and her expenditures on schooling ($P < 0.01$). Sharing income increases with the age of the female, implying that with experienced or older women, men are more willing to share their income with their wives. This result may be explained by the fact that mutual trust grows with the duration of the spouses' bond. The wife's ownership of small ruminants induces more sharing of income from husband, because sheep and goats constitute a kind of bargaining power for women allowing them to receive more compensation which allows for more income and economic power. The husband's income has a positive effect on 'sharing income', suggesting that wealthier husbands are more willing to share their income with their wives. In contrast, cash transfers from migrants allowed to the wife have a decreasing effect on 'husband sharing income'. This means that husbands are unlikely to share their income when their wives receive cash transfers. The wife's contribution to schooling expenditures also had a positive effect on 'sharing factor'suggesting that the more responsibility the wife takes in the public sphere, the more the male is likely to share income, suggesting a kind of collaboration in making expenditures. Ethnicity further explains the factor 'sharing income', and Wolof and Peulh are less likely to share their income than the Sereer (our reference group). These results may be explained by the fact that since households in the AP area (mostly Wolof) earn less, partners are more likely to give priority to basic needs (food, schooling), leaving little to share. The adjusted R^2 equals 0.32 and we can strongly reject the null hypothesis that the coefficients are equal to zero at ($P < 0.001$). Unless for the ethnicity and area which show large beta coefficients (0.37 for Wolof group, 0.34 for peulh group and 0.37 for area), most variables in the regression analysis of the 'sharing factor' show standardized coefficients that are not very meaningful from an economic point of view (for husband's income, the standardized coefficient = 0.18 and for the wife's assets in livestock the standardized coefficient = 0.12). This means that belonging to a particularly group of people or community may have an economic meaning with respect to the 'sharing income factor'.

Factor 3 (borrowing food or money) captures responsibility for food provision and appears to be associated with the female's age ($P < 0.01$). This means that experienced women are more responsible for borrowing money or food to fulfil household needs. This result may explain that the oldest women have more power in the household when it comes to responsibility for food provision. Husbands' off-farm wages have a decreasing effect on the factor 'women borrow food or money,' suggesting that in particular males'activities beyond the household, allow for more security in food provision ($P < 0.01$). The wife's responsibility in borrowing food and money for the household decreases with her land ownership, suggesting that physical assets is associated with less borrowing food or money (probably when women have access to land, they contribute to food provision by purchasing more and borrowing money or food less). A similar effect was found for the wife's access to credit which also had a negative effect on the factor 'borrowing' ($P < 0.01$), suggesting more power in the provision of food (purchasing more and borrowing less). The R^2 of the regression equals 0.17. The Wald test shows that the null hypothesis is strongly rejected at $P < 0.001$. We computed the beta coefficients some of which show significant economic effects besides statistical significance. An important point is that wife's access to credit and land lead to the fact that she has less to borrow for food or to borrow money to buy food (the standardized coefficients are respectively 0.24 and 0.34). This effect has an economic implication on the poverty issue and the vulnerability of some households.

In the fourth regression analysis concerning the factor 'management of finance,' the estimation shows that the wife's responsibility for the management of finance increases with her contribution to expenditures (food at $P < 0.05$ and health at $P < 0.01$). Another finding is that the wife's responsibility for finance management decreases with the husband's income and the wife's age, suggesting that when the husband's income increases, he makes most of the decisions related to the management of finance, in contrast to couples making joint decisions, The latter finding would be in line with other findings in Senegal which underlined that when the husband's revenues increase, women tend to have a role more complementary to men's, and have less responsibilities in decision-making, while increasing their own capital (resource accumulation, social capital) (Ndiaye Diop, 2007). Household size also has a negative effect on the women's management of finance, suggesting that in large households men tend to be exclusively responsible for the management of the household income. For religion, we notice a decreasing effect on the management of finance ($P < 0.01$): Muslim wives seem to manage finances less than catholic wives. In fact, Islam as the major religion in Senegal designates the male head of the household as the one responsible for it. Consequently Islam favours a social and cultural environment where women are constrained in taking responsibility within the household and the community (Amin, 1997; Ghuman, 2003; Hochschild, 1991; Pyke, 1996). From this point of view, gender roles are reinforced by social norms imposing specialization in specific activities for men and others for women. For example, as in many developing countries, Muslim women are assigned home tasks and men will do market work (Caldwell and Caldwell, 1993). This situation contrasts with Christian women who are more willing to bargain with their partners. The adjusted R^2 in the estimation model equals 0.29 and the Wald test rejects the null hypothesis at ($P < 0.001$). The computed standardized coefficients show acceptable values for monthly husband's income and zone. These values are respectively 0.21 and 0.23 suggesting an economically meaningful effect on the factor 'management of finance.' For example, to the extent that husband's income may reduce women's management of finance, one may expect some substantial negative consequence on household welfare (because women have a positive contribution on household-wellbeing).

4.4 Conclusion

Household food is provided through farm production and market activity. Results concerning the responsibility for meeting household needs, show that more than wives, husbands are responsible for providing household goods and managing finance. The responsibility for food provision also includes borrowing money to buy food (77% of total cases), borrowing food (68% of total cases) or bartering household assets for food (28% of total cases) with relatives or neighbors. A comparison between zones shows that women in the SP area have more voice in decisions concerning the management of finance, the purchase of food and loans for buying food. Decisions on health (type of treatment, use of medical care) mainly depend on the husband; the wife intervenes only in a few cases. The responsibility for financing health care and medical treatment depends on the budget share of the husband who is the main breadwinner in the household (the husband generally decides on which type of health problem the money should be spent). In some cases the recourse to traditional healers prevents the use of modern medicine. This situation leads to less than full use of medical care and other health facilities, and increases the risk to vulnerable people (women and children) of high morbidity and catching serious diseases. Results show that this situation is often a consequence of the high cost and lack of knowledge of modern treatment. In addition, a substantial part of the sample does not usually discuss family planning and sexually transmitted diseases. When it comes to decisions on caring for ill members, women are by far the major providers of this household service. In fact it is an additional burden to women that increases their family responsibility and limits their participation in the labour market.

Women's power measured through factor analysis, shows that important proxies of their relative status in the household are those related to their mobility, their decision-making in food provision and income management, and their willingness to share their income with their husbands and vice versa. The mobility factor is related to women's autonomy in access to health care and health facilities and to the purchase of modern medicine.

The results of the four regressions regarding the determinants of women's power suggest that the status of women is associated not only with their labour income, but also with their age and other indices of bargaining power, such as their contribution to expenditures (especially household goods like food and schooling), their savings (non-labour income), assets (goats and sheep) and remittances. The responsibility of women for the management of the household as reflected in their contribution to health, school or food expenditures, are consistent with most empirical evidence (Doss, 1996; Duncan and Chen, 1994; Quisumbing and Maluccio, 2000; Abadian, 1996; Jacobson, 1992). However, it appears that in decision-making, Muslim wives are constrained through binding norms. While the education of the partners has no effect in our study, evidence from other studies showed that human capital is very important in intrahousehold decision-making (Lawrence *et al.*, 2007). For example, recent studies conducted in Ghana (Antwi-Nsiah, 1993) show that educated households generally made more joint decisions. The lack of effect of the variable education in our empirical result may be related to the generally low level of education in the two areas, especially among women. In addition, it seems that in wealthier households where the husband's income is determinant, women tend to have less voice in the management of finance and thus in decision-making. Another finding is also that some important variables (wife's access to credit, wife's access to land, husband's income, ethnicity and area) have significant standardized coefficients that suggest that the effects depicted have an economically meaningful implication.

The following chapter will give more insight into household decision-making on consumption expenditures. For example, how do earnings from husbands and wives con-tribute to consumption? Are spouses' incomes pooled? How do husbands and wives contribute and behave towards expenditures?

Chapter 5

Bargaining over expenditures and gender utility functions

5.1 Introduction and theory

5.1.1 Introduction

Having described in Chapter 4 how household decisions are made regarding food provision, health care and treatment, Chapter 5 is related to the pattern of consumption and to the way husband and wife behave regarding expenditures. In this chapter, we provide empirical evidence regarding two matters: (1) whether in the process of spending on household goods, income is pooled and (2) the estimation of utility derived from the consumption of household goods versus private expenditures. This chapter contributes to the issue of husbands' and wives' individual interests in relation to the welfare and the well-being of the household. The analysis is based on a description of the way in which husband and wife contribute to the household consumption pattern, and on modeling the bargaining relationship between husband and wife concerning the allocation of income.

5.1.2 Models of household decision-making: theory and relevant findings

Household behaviour combines the interests of its members. A theoretical issue concerns the impact of individual sources of income (husband's or wife's) on the pattern of household expenditures. In fact, the issue is how the income earned by the husband and the wife separately is spent on various consumption categories. Samuelson (1956) assumed a household utility function capturing common preferences, reached by consensus among household members. Becker's unitary household model (1974, 1981) assumes that family expenditures are dictated by one person (usually the husband) who takes into account the other members' preferences.

Even though Becker's model still is a strong approach to household decision-making, the model has been criticized and rejected by many researchers in many countries (Horney and McElroy, 1988; Browning and Chiappori, 1998; Strauss *et al.*, 1995; Haddad *et al.*, 1997). The application of the unitary model of the household has led to several policy failures in agriculture, a critical one being that targeting one spouse (usually the husband), rather than both spouses, led to the non-adoption of particular policies or unexpected policy outcomes (Browning and Chiappori, 1998, Lundberg *et al.*, 1997). For example, rice-growing technologies in some West African countries are addressed to the household heads, leaving out the workers in the fields, namely women (Elson, 1992). Alternative models (Chiappori, 1992) assume that female and male income could be allocated differently, implying that the utility of consumption may differ between spouses. In these models the household maximizes a collective utility function with the exercise of some power by one partner on the other (Apps and Rees, 1997; Browning and Chiappori, 1998). These models can, for example, imply 'Pareto-efficient models' which allow for an efficient allocation of resources; this means that each spouse chooses labour supply and consumption to maximize his or her utility (Chiappori, 1992).

Previously, it was established that husband and wife in West-African countries tend to make consensual decisions on expenditures and resource allocation (Guyer, 1981; Saul, 1989; Smith and Chavas, 1999). There have been some empirical attempts at assessing the household decision-making process in developing countries. Most of them assume that income is pooled and spent in a consensual way implying Pareto efficiency (for example, after reallocation of income, at the optimum no member can be better off without the others being worse off). However, research has provided empirical evidence for the failure of Pareto efficiency in intrahousehold decision-making (Doss, 1996; Udry, 1995, 1996). Also, women and men tend to make spending decisions separately and individually rather than on the basis of 'pooled' income (Whitehead, 1990; Doss, 1997). The latter approach implies a combination of household decisions, some of which can be made jointly, some separately (Carter and Katz, 1997). Until recently, the standard of thinking of both theoretical and empirical analysis was a 'common preference' model of the family which assumes that family members act as one person. However, studies in African countries have shown conflict between spouses concerning the control over cash-crop earnings and labour-force contribution (for example, the case of cotton in Burkina Faso, rice in Cameroun and coffee in Ethiopia, see Lim Sung *et al.*, 2007; Brown and Haddad, 1996; Haddad *et al.*, 1997; Smith and Chavas, 1999). In fact the greater individual access to earnings shows the partners' exercise of bargaining power (Lilja *et al.*, 1996, Sen, 1990; Doss *et al.*, 2009). These findings suggest gender-specific roles in expenditures and responsibilities, and differences in spending preferences. Nowadays, bargaining models predict that, except for different utility functions, the differences between men and women are found in their bargaining power: women may receive more utility through bargaining instead of accepting male decisions, suggesting a sort of conflicted cooperation (Himmelweit, 2001; Sen, 1990). This argument is used in collective models leading to either non-cooperative or cooperative models (Lundberg and Pollak, 1993; Kanbur, 1991; Vermeulen, 2002). Cooperative models are based on communication and reaching binding agreements on the distribution of the household benefits: agreements that are enforceable at no cost. Non-cooperative models are based on 'best responses' from each partner to the other player's strategies (Lundberg and Pollak, 1994). One group of non-cooperative models includes pure non-cooperation where each partner optimizes his or her utility by taking the other's behaviour as given (Browning, 2000; Chen and Voolley, 1999). The bargaining relationship between husband and wife is an important determinant of the control over income, access to food security and good health. (Carter and Katz, 1997; Himmelweit *et al.*, 2001, Lim Sung *et al.*, 2007). For example, in the context of areas where livestock is an important production factor, the wife's responsibility in dairy production co-determines her power in bargaining with her husband because she controls the earnings from milk production. On the other hand, the man's control over land co-determines his control over income from selling livestock and other, crop-related, activities.

In bargaining models it is assumed that preferences of household members may differ, implying preference heterogeneity in household decision-making and separate budget constraints. Under preference heterogeneity, there are several models of household behaviour: (1) cooperative bargaining models (Lundberg and Pollak, 1993; McElroy and Horney, 1981; McElroy, 1990), (2) non-cooperative models (Lundberg and Pollak, 1994) and (3) the unified household model (Chiappori, 1988; Browning and Chiappori, 1994). The latter model assumes that couples jointly choose a Pareto-efficient outcome. Among the bargaining models (cooperative or non-cooperative), the Nash bargaining model is the best-known cooperative household model. It has three components (Pollak, 2005): (1) a feasible set in the utility space, (2) reservation utilities for each family member, and (3) a threat point that reflects family members' bargaining power. In theory, the equilibrium values are the utilities associated with a non-cooperative equilibrium within the marriage. For example, in the Nash bargaining model (Nash, 1950; 1953) the utility received by husband or wife in the solution, depends upon the threat point: the higher one's utility at the threat point, the higher one's utility in the Nash bargaining solution will be (Lundberg and Pollak, 1996). The Nash bargaining model implies that partners agree on strategies before engaging in household decision-making, suggesting binding and self-enforcing agreements. Non-cooperative models, on the other hand, do not assume that husband and wife use binding norms (Lundberg and Pollak, 1994), or agreements enforceable at no cost, but rather imply that each partner develops his or her own strategy, taking the behaviour of the other partner as given. The non-cooperative model leads to equilibrium such that each partner's strategy is the 'best' response to the other partner's strategy. In addition, the non-cooperative model neither assumes, nor rejects Pareto efficiency.

Despite the remarkable contribution of bargaining models in the theoretical reflection on household decision-making, the debate regarding the choice of the model remains controversial because according to Lundberg and Pollak (1996):no new framework has gained general acceptance as a replacement of the common preference model. Also, consumption decision-making regarding the way in which income should be spent is mostly controversial in developing countries. For this reason we start our analysis by looking whether or not husband and wife make decision jointly or not. Because of social and cultural gender roles, decision-making on the way that income should be spent may be taken unilaterally by the husband since he is the major provider of goods in the household. In Senegal for instance, because 90% of people are Muslim and given the influence of this religion on the cultural behavior of household members, women would be marginalized in taking decisions on the way of spending income. Since consumption decision-making co-determines on the well-being of the household members, we are interested in the first part of this chapter in understanding whether or not income is pooled in spending on consumption. We expect that despite the fact that some cultural and religious factors may negatively impact on their power in decision-making, women in rural Senegalese households are somehow independent in the way their income would be spent.

In the context of Sub-Saharan African countries, particularly in Senegal, as we showed in Chapter 3, males and females earn their incomes separately (these differ across zones), and each partner contributes to the household activities and expenditures, attempting to achieve food security and meet household needs. Hence, our main hypothesis is that in both areas, income initially is not pooled in the household, and that husband and wife engage in a sort of cooperation to attain the optimal level of well-being and food provision for the household. In order to capture the interdependence between husband and wife in the provision of goods, and to measure the degree of interdependence in the household, we use two different and complementary approaches. In the first we describe the participation of each partner in making expenditures, and we use Engel curves to demonstrate whether or not income is pooled. This first approach is important to deal with because whether income is pooled or not will help in supporting and choosing our household decision-making model (common preferences model, cooperative bargaining model or non-cooperative bargaining model). In the second approach, we assume a non-cooperative bargaining model without pooling. However, bargaining is a form of distributive negotiation process that is both competitive and positional and it leads to a decision-making for the negotiation of an agreement. Hence, bargaining in consumption decision-making is a dynamic process. In our study, since we do not have available panel data on consumption, we will focus in this chapter on modeling household decision-making. In this model the equilibrium solution is the result of bargaining that maximizes the value that husband and wife attach to the consumption of private and household goods.

5.2 Modeling household expenditures and test of income pooling

The data concerns 300 rural dual-headed households whose husbands and wives are married couples whose income stems primarily from agricultural activities and secondly from non-agricultural activities or transfers (see Chapter 3). Separately, husband and wife list and report monthly expenditures on major consumption goods and other household needs (cereals, oil, rice, sugar, fish, tea/coffee and ingredients, health, clothes and schooling). These expenditures have been included in the estimation of Engel curves. In this approach, we test how incomes are used by husbands and wives and whether any additional CFA franc earned by the husband will be spent in the same way as any additional franc earned by the wife. In doing so, we test the hypothesis of income pooling by using Engel curves.

In the literature, pooling income has different meanings. For example, to sociologists, pooling refers to the way couples manage their money (Pahl, 1983; Treas, 1993; Woolley, 2003); whether they have just one bank account, or two or more accounts to manage their finances. To economic psychologists, pooling income may refer to the decision-making process in the management of finance (Bonke and Browning, 2009). For economists, pooling is a property of the demand function and refers to the fact that a transfer of money from husband to wife or from wife to husband will not affect the expenditure pattern (Pollak, 2005). The latter definition reflects our interest, and suggests that economists study how additional income affects power and the distribution of consumption. Then, there is a strong link between decision-making power and pooling income from an economic point of view. Since demand for goods depends on prices and total income, Engel curves demonstrate the dependence of expenditures, E, on the sum of husbands' and wives' incomes, $Y_h + Y_w$, holding prices and other demographic characteristics, Δ, constant.

$$E = f(Y_w + Y_h | \Delta) \tag{5.1}$$

Equation 5.1 is a restriction of the more general Engel curve

$$E = f(Y_w, Y_h | \Delta) \tag{5.2}$$

Equation 5.3 is a general specification of Equation 5.2 that can be attributed to any collective model of household behavior and contains a quadratic specification of the Engel curve:

$$E = \beta_o + \beta_w Y_w + \beta_h Y_h + b_{ww} Y^2{}_w + \beta_{hh} Y^2{}_h + \beta_{wh} Y_w Y_h + \textstyle\sum_i b_i \Delta_i + e \tag{5.3}$$

Δ_i, $i = 1\ldots, n$ are demographic variables reported, such as the size of the household, area, age of husband and wife, religion, marital status (polygamous household or not), transfers from migrants, ethnicity and the wife's access to credit and savings.

Household size may explain the expenditure for a particular household, because it reflects the volume of the demand for goods. Differences between areas in terms of accessibility to markets, availability of infrastructure (roads, type of transportation) and behaviour, may affect the availability of facilities not too far away (for instance school, health care in a hospital or health centre, etc.). We expect that the level of education may influence the expenditures on some facilities (i.e. school attendance or professional health care). The wife's savings (capital accumulation from livestock, trade activities or social-capital relations) and the wife's access to credit, could be good indicators of her share in expenditure on food provision, health care and participation in children's schooling. We expect that the wife's savings will increase with her contribution to expenditures. Religion and ethnicity which impact on customs and cultural behaviour, may be associated with the consumption of certain kinds of services and facilities instead of others. Finally, we may expect that women in polygamous households may reduce their participation in the purchase of food products, and increase their consumption of private goods (health, clothing).

The term e in Equation 5.3 is a random error term. We assume that income pooling results in restrictions on Equation 5.2 such that $\beta_w = \beta_h$ and $\beta_{ww} = \beta_{hh} = \beta_{hw}/2$ meaning that each income (from husband and wife) has the same effect on expenditures. Equation 5.2 then can be re-written in a restricted form giving:

$$E= \beta_0 +\beta_1(Y_w+Y_h)+ \beta_2(Y_h^2+Y_w^2+2Y_hY_w) +\sum_i b_i\Delta_i+ e \quad (5.4)$$

The rejection of these restrictions implies that income is not pooled. The restrictions were tested for all expenditures on major consumption goods in the household. The income of husband and wife corresponds to annual revenues from livestock, agricultural and non-agricultural activities, and net of costs. Costs are related to expenditures on inputs: fertilizers, seeds, water, animals, health, feed and other costs. Expenditures were considered per month for each need regarding food, and per annum for schooling and health care. We investigated the variation of expenditures with respect to husband's and wife's income by using OLS.

5.2.1 Household expenditures on food and non-food consumption goods

In rural Senegalese households, goods produced for consumption are cereals, milk and oil (traditionally processed from the available groundnut). The main consumption goods are cereals, milk and rice (the latter in the north and the south of Senegal where rice is cultivated). But with the change in patterns of consumption (due to imported goods such as rice, powdered milk and others), rural households tend to behave like urban consumers; for example consuming more rice and other imported goods (tea, coffee and sugar). Table 5.1 shows that for the head of the household, the husband, cereals have top priority among household needs. Next come oil, sugar, rice and fish. Bread, milk and meat are given less priority.

Even when most households grow food crops, mainly millet, for their own use, cereals constitute a major part of expenditures on consumption goods. In fact, a substantial part of the millet is sold on the market, but until the end of the dry season, households are frequently forced to buy millet back from the market, sometimes at a higher price. From our data, estimations show that only 10% of the households do

Table 5.1. Average expenditures (CFA)[1] on consumption goods.

Goods	Husband		Wife		Household, total sample		Household, SP area		Household, AP area	
	Mean	Std	Mean	Std	Mean	Std	Mean	Std	Mean	Std
Food (monthly)										
Cereals	19,309**	1,208	1,617	372	37,904	2,247	33,114	2,580	45,051**	4,101
Oil	13,127**	657	2,806	339	17,191	37	18,321	925	16,179	9,743
Sugar	9,317**	477	1,822	297	11,538	570	12,864**	775	9,163	697
Fish	5,833**	822	3,923	637	11,583	748	13,396***	2,036	10,777	578
Rice	20,109**	2,360	1,659	273	5,169	575	6,066	853	4,278	768
Tea and coffee	7,410**	829	1,348	332	8,702	915	8,732***	977	8,440	2,695
Milk	2,120	1,021	272	272	1,477	645	210*	210	9,500	3,041
Ingredients (vegetables, spices, peanut)	6,777**	833	2,450	449	12,000	874	13,200***	4,683	12,000	875
Non-food (yearly)										
Schooling	33,184	4,804	5,711	1,620	44,300	5,633	33,180*	6,094	55,265	9,375
Health	36,331	3,254	6,638	1000	48,134	3,726	56,735*	6,161	39,647	4,130
Clothes	40,000*	2,639	26,346	1,960	74,260	3,689	82,355*	5,597	66,006	4,745

[1] 1 US$ = 475 CFA during the year of the survey.

* $P<0.05$, ** $P<0.01$, *** $P<0.001$ (differences between husband and wife, and areas, respectively).

not buy cereals and keep all the household production for their own consumption. Rice, which according to farmers is a well appreciated food, is purchased by 34% of the households. In fact, a change in farmers' attitudes to rice has rendered it more important than cereals in daily household consumption. Similar attitudes can be noticed for coffee, tea and oil.

The results show that women contribute to all household expenditures, thus highlighting their contribution to food security and welfare of the household (Table 5.1 and Table 5.2). For example, wives contribute 34% of household spending on fish, 32% on rice, 20% on ingredients (vegetables, peanut, spices, tomato and others), 16% on oil, 16% on tea/coffee and 4% on cereals. Other important contributions from women related to schooling, clothes and health expenditures, are also reported, amounting to 13, 36 and 14%, respectively. These results show the role of women in decision-making and in the provision of consumption goods.

A comparison between areas shows that, in general, SP households spend more on food (rice, sugar, fish, oil and ingredients), though cereals are purchased more in the AP area. This can be explained by two possible effects brought out during our personal interviews: SP households keep all their home-grown millet for their own use or compensate by consuming a lot of milk. Comparison across zones, of women's

Table 5.2. Wife's contribution to expenditures (in CFA)[1] by area.

	AP area (N=151)		SP area (N=149)	
	Mean	Std	Mean	Std
Tea and coffee	-	-	878***	222
Ingredients	1,801	298	442**	309
Milk	2,000	200	-	-
Fish	1,158	264	879	280
Sugar	537	162	1,742**	343
Cereals	652	309	109	271
Oil	892	230	3,917***	517
Rice	417	200	2,907***	489
Total share in food	5,459	806	11,865***	1,319
Schooling	4,357	1,844	7,083	2,675
Clothing	15,947	1,529	36,885***	3,427
Health	1,976	544	11,395***	1,846

[1] 1 US$ = 475 CFA during the year of the survey.
** $P<0.01$, *** $P<0.001$, (-) no data available.

contributions to expenditures shows that SP wives contribute more than AP wives (Table 5.2) to all needs including clothes, health and schooling. The differences in contributions are significant for most needs (Table 5.2). These results suggest that SP women are economically more powerful than AP women. Moreover, economic power can explain their more intensive participation in decision-making and contribution to expenditures.

5.2.2 The test of income pooling

For all goods considered, we test whether the estimates from OLS show that the restrictions on the Engel curves apply i.e. that $\beta_w = \beta_h$ and $\beta_{ww} = \beta_{hh} = \beta_{hw}/2$, see Table 5.3. To test this restriction, we use the Likelihood-ratio test for each estimated Engel curve. To perform the Likelihood-ratio test, we use Equation 5.4 as a restricted version of Equation 5.3. The likelihood function is:

$$L(e)=\prod \frac{1}{\sigma\sqrt{2\pi}} e^{\frac{-e^2}{2\sigma^2}} \quad (5.5)$$

With $\prod$ being the multiplication operator multiplying the probability of all observations.

If we take the logarithm of the function, it becomes:

$$\log L(e)=\sum(-\log \sigma\sqrt{2\pi} - \frac{e^2}{2\sigma^2}) \quad (5.6)$$

e = error term, σ^2 = variance of e.

To test our hypothesis of income pooling, the statistic *-2(LLU-LLR)*, which is Chi-square distributed with as many degrees of freedom as the number of restrictions, should not be significant (with *LLU* = log-likelihood of the unrestricted equation; *LLR* = log-likelihood of the restricted equation).

To make more precision on our results, we estimate two regressions: one with the quadratic equation (Table 5.3) using Equation 5.3 in order to test for $\beta_w = \beta_h$ and $\beta_{ww} = \beta_{hh} = \beta_{hw}/2$ and the other using a Cobb-Douglas equation in order to capture the variation of income with respect to expenditures (including only husband's and wife's income and other socio-economic factors that may also impact on expenditures) (Table 5.4). The results show that the rejection of the null hypothesis of income pooling is binding for tea/coffee ($P < 0.05$), clothes ($P < 0.05$) and health ($P < 0.01$) (Table 5.5). For all other foods and schooling, we cannot reject income pooling. Surprisingly, in Table 5.3, expenditures on food vary only with monthly husband's income and Table 5.5 shows that income is pooled to be spent in a consual way for this good. This may be explained by the fact that the difference between the coefficients β_w and β_h are so significantly high to explain divergence in spending on food. For tea and coffee, Table 5.3 does not show variation of expenditures with

Table 5.3. Estimates of the Engel curves in the total sample.

	Foods (cereals, oil, sugar, fish, ingredients)		Tea and coffee		Health		Clothing		Schooling	
	Coefficient	t-value	Coefficient	t-value	Coefficient	t-value	Coefficient	t-value	Coefficient	t-value
Monthly husband's income squared	-1.04×10^{-7}	-1.48	-1.67×10^{-8}	-1.53	-1.55×10^{-8}	-1.75	-9.43×10^{-9}	-1.18	-3.46×10^{-10}	-0.03
Monthly wife's income squared	-5.43×10^{-7}	-0.68	-1.01×10^{-7}	-0.81	-3.25×10^{-7}	-3.20**	2.80×10^{-7}	3.05**	-2.43×10^{-7}	-1.56
Monthly husband's times monthly wife's income	-2.00×10^{-7}	-0.43	1.19×10^{-7}	1.66	2.16×10^{-7}	3.72***	-1.63×10^{-7}	-3.11**	-2.90×10^{-8}	-0.33
Monthly husband's income	0.110	2.58**	0.007	1.05	0.011	2.04*	0.026	5.48***	0.005	0.72
Monthly wife's income	0.187	1.28	-0.040	-1.80	0.029	1.61	-0.008	-0.50	0.079	2.79**
Adjusted R^2	0.08		0.07		0.18		0.31		0.13	

* $P<0.05$, ** $P<0.01$, *** $P<0.001$.

Table 5.4. Determinants of husband's and wife's expenditures.

	Foods (cereals, oil, sugar, fish, ingredients)		Tea and coffee		Health		Clothing		Schooling	
	Coefficient	t-value	Coefficient	t-value	Coefficient	t-value	Coefficient	t-value	Coefficient	t-value
Monthly husband's income	0.0451	2.15*	-0.0172	-0.15	0.1783	2.04*	0.0877	3.36**	0.1721	1.81
Monthly wife's income	0.0385	3.92***	-0.1710	-3.25**	0.0728	1.77	0.0097	0.79	0.1414	3.16**
Age of husband	0.0003	0.06	-0.0324	-1.14	0.0155	0.70	0.0166	2.50**	0.0060	0.25
Age of wife	-0.0075	-1.30	0.0227	0.78	-0.0382	-1.68	-0.0108	-1.59	0.0407	1.64
Zone [1]	0.1300	1.16	-3.9797	-6.66***	-0.4163	-0.89	-0.4933	-3.54***	0.9418	1.85
Religion [4]	-0.0162	-0.07	3.0896	2.50**	-1.958	-2.03*	-0.1284	-0.44	-0.1793	-0.17
Polygamy [5]	0.0073	0.24	-0.0387	-0.24	0.0362	0.28	0.0363	0.95	0.0711	0.51
Household size	0.0245	2.60**	0.06703	1.33	0.0832	2.11*	0.0420	3.57***	0.1336	3.11**
Wife's savings	1.99e-07	1.29	1.08e-06	1.31	1.27e-06	1.98*	5.37e-07	2.79**	1.07e-06	1.53
Wife's access to credit [3]	-0.0066	-0.07	0.5665	1.14	0.6442	1.66	0.0798	0.69	0.2573	0.61
Wolof [6]	0.2147	1.67	-1.4702	-2.14*	-0.2908	-0.54	-0.3133	-1.95*	-0.2548	-0.44
Peulh [7]	0.2962	2.12*	0.2252	0.30	-1.180	-2.02*	-0.4203	-2.41**	-1.127	-1.77
Husband's education [3]	-0.0856	-0.97	-0.1307	-0.28	-0.1451	-0.39	0.0968	0.88	0.9138	2.28*
Wife's education [2]	-0.0992	-1.00	-0.2529	-0.48	-0.3152	-0.76	0.2130	1.72	0.8138	1.80
Constant	9.651	17.33	3.875	1.30	8.5546	3.68	7.5828	10.92	-1.569	-0.62
Wald test for the rejection of the null hypothesis	F(4, 270)=9***		F(4, 270)=13,48***		F(5, 270)=4.48***		F (7, 270)=7.30***		F (3, 270)=8.59***	
Adjusted R^2	0.12		0.32		0.17		0.15		0.24	

* *P*<0.05, ** *P*<0.01, *** *P*<0.001.

[1] Dummy (0 = SP area, 1 = AP area).

[2] Dummy (0 = lowest, 1 = highest).

[3] Dummy (1 = yes).

[4] Dummy(0 = Catholic, 1 = Muslim,).

[5] Dummy (1 = polygamous household).

[6] Dummy Wolof (Wolof = 1, Sereer = 0).

[7] DummyPeulh (Peulh = 1, Sereer = 0).

Table 5.5. Likelihood ratio test for the restriction that $\beta_w=\beta_h$ and $\beta_{ww}=\beta_{hh}=\beta_{hw}/2$ in the total sample and by area.

	Total sample N=300	SP area N=149	AP area N=151
Expenditures on food	2.84	2.92	1.39
Tea and coffee expenditures	9.62**	6.84*	2.78
Health expenditures	14.20***	11.68***	5.93
Clothing expenditures	8.52**	6.84*	4.17
Schooling expenditures	5.68	1.46	82.40****

*$P<0.10$; **$P<0.05$; ***$P<0.01$; **** $P<0.001$.

respect to husband's and wife's income. However, Table 5.5 shows the rejection of pooling for this particular good that can be explained through probable significant difference between the coefficients β_{ww} and $\beta_{hw}/2$ and between β_{hh} and $\beta_{hw}/2$. The same reasoning can be applied to schooling, differences between the two coefficients β_w and β_h are not significant in indicating no income pooling. These results suggest that husband and wife may not pool their income in spending on health, clothing, tea and coffee, but probably do pool their incomes in paying for food and schooling. Looking at the Likelihood-ratio tests in the two areas, the results show that in the SP area, income is not pooled for expenditures on health ($P<0.01$), clothes ($P<0.1$) and tea and coffee ($P<0.1$). However, in the AP area, the rejection of income pooling holds only for schooling ($P<0.001$) (Table 5.5).

These results reject the unitary model of the household which assumes one household utility function. This implies that husband and wife behave and react differently with respect to expenditures. In many poor countries, it seems that access to health care and health facilities may be difficult because of the related costs and the lack of infrastructure in these countries. Also, it can be said that only wealthier people have access to valuable clothes. And husband and wife set priorities: first come primary household needs, i.e. food; next, if income suffices, they deal with health costs and private goods (clothes). This may explain that husband and wife obviously do not pool their income to spend on clothes and health that are generally expensive. The same can be said for tea and coffee which depend on individual needs. It appeared that the expenditures on food are handled in a consensual way in order to make ends meet. However, husband's and wife's incomes are spent differently on health, clothing, tea and coffee. The rejection of the unitary model in our study is consistent with most empirical findings in developing countries (Duncan, 1996; Quisumbing, 1994; Quisumbing and Maluccio, 2000).

In order to express the relationship between the partners' incomes and household expenditures in terms of elasticities, estimations are made using a Cobb-Douglas equation using their respective logarithmic values (Table 5.4). The results show that expenditures on clothes and health vary significantly with the husband's income ($P<0.01$ and $P<0.05$, respectively). One percent increase in the husband's income leads to an 8% increase in clothing expenditures and 17% increase in health expenditures, suggesting that these needs are in his sphere. In fact, since the husband generally earns more than the wife, he may be in charge of these expensive goods. On the other hand, expenditures related to schooling vary significantly only with the wife's income ($P<0.01$); and are more elastic with her income, suggesting that women's additional gains are likely to be spent on schooling. Nowadays, in many African societies women more than men care about their long-term children's future, especially that of girls. Also, evidence showed that when the wife's assets increase, school attendance of girls increases (Gallasso, 1999). Because to some extent daughters replace their mothers in household tasks, one may expect that the decision to motivate them towards a better education depends on their mother's involvement in decisions and her financial involvement in schooling. Moreover, the gender gap in productivity has been mostly attributed to women's lower education, which is why much effort has been spent on girls' education in developing countries (World Bank, 1993; World Bank, 2004; Lewis *et al.*, 2008; Lopez-Claros *et al.*, 2005; Lifanda, 2005), and in many countries women are personally responsible for their daughters' schooling.

Food expenditures increase significantly both with the husband's and the wife's income (respectively, $P<0.05$ and $P<0.001$); 1% increase in the husband's income leads to a 0.45% rise in food expenditures and for the wife, the expected rise is 0.38%. In fact, expenditures are made jointly in the way that food provision is handled by the couple, both contributing at different levels. The income-pooling hypothesis for food was not rejected, suggesting that couples have a consensual way of spending on food. In fact, in West-African countries, food expenditures are generally the responsibility of the husband who is the chief of the household. But wives contribute substantially, depending on their share of income and assets (IFPRI, 2000; Datt *et al.*, 1998; Quisumbing *et al.*, 2000; Maurique *et al.*, 2008). Also, because of the important role women play in social-capital accumulation (quality of an individual's link to other individuals and other groups, involvement in groups of interest), they are able to participate in food security and poverty alleviation to the same extent as men (Cross, 1999). Moreover, women's capital accumulation through the informal market (friends, neighbors and relatives) via free-borrowing is by far the most used financial instrument in developing countries to combat poverty and increase small savings (Collins, 2009).

A comparison between areas (Appendix 3) shows that expenditures on food in the SP area vary significantly only with women's income ($P<0.001$), and in terms of income elasticity, more so than with men's income (0.069 compared to 0.037). This finding suggests that the wife's income has a powerful effect on food provision in the SP area. The latter finding may be explained firstly by the economic position of women in this area, and secondly by the absence of men when they are looking for pasture for their livestock. Regarding clothes, AP expenditures vary only with the wife's income. For health and schooling, in the SP area positive significant elasticities are noticed for the wife's income: 1% increase in the SP wife's income leads to 19% and 27% increases of health and schooling expenditures, respectively. These findings recognize that SP wives are more willing to spend their earnings on the well-being of the household than AP wives. While the wife's education significantly increases expenditures on clothes in the AP area, the SP husband's education leads to a decrease of expenditures on food. This finding suggests that with higher education, males tend to shift their consumption pattern from food to other goods (perhaps to other unknown private goods). Large household size raises expenditures on all goods in both areas.

5.2.3 Determinants of expenditures

Equation 2 is explained by individual and community factors. The 'goodness of fit' measure (adjusted R^2) was 0.12 for food, 0.32 for tea and coffee, 0.07 for health, 0.15 for clothes and 0.24 for schooling (Table 5.4).

Important determinants of expenditures are zone, household size, ethnicity and wife's savings. The size of the household predicts expenditures on food ($P<0.01$), health ($P<0.05$), clothes ($P<0.001$) and schooling ($P<0.01$). All estimates are positive. One percent increase in household size induces an increase of expenditures on food (2.40%), health (8.10%), clothes (4.20%) and schooling (13%).

The area of living and ethnicity is associated with the patterns of expenditures regarding tea/coffee and clothes. SP expenditures on tea/coffee and clothes are significantly lower (both $P<0.001$) than in the AP area. There, expenditures on tea/coffee and clothes decrease by 4 and 5%, respectively. In fact, consumption depends on cultural and socio-economic characteristics. The effect of lower clothing expenditures in the AP area may be explained by the fact that the SP area is economically more comfortable than the AP area due to assets and earnings. Wife's savings positively and significantly predict health and clothing expenditures ($P<0.05$ and $P<0.01$, respectively). This finding suggests that wife's savings reinforce the household's capability of buying clothing and spending on health. And this finding demonstrates that the wife's savings due to livelihood diversification (generally livestock accumulation in the SP area, and revenues from trade in the AP area) or social-capital endowment (rotated financial help from friends), serves as a kind of wealth accumulation used for the well-being of the household. Many studies in developing countries have shown the link between savings and recovery from shocks such as medical treatment or death following illness (Smith, 1999; Lundberg, 2003; Kong *et al.*, 2005). This result is in line with other findings (e.g. Duncan, 1990) in Brazil showing that children did better in terms of mortality and morbidity when their mothers controlled a larger fraction of the couple's non-labour income. Other important predictors of consumption are ethnicity and religion. Belonging to the Peulh is positively associated with food expenditures ($P < 0.05$) but negatively associated with health expenditures ($P < 0.05$) and clothing expenditures ($P < 0.01$). This finding suggests that the relatively wealthy Peulh have larger expenditures on food but are unlikely to spend their additional income on health and private goods. Also, private goods expenditures (on clothing, etc.) are significantly lower in Wolof group compare to the Sereer group suggesting that to them clothes are not priorities.

As for religion, we observed significantly higher expenditures on tea and coffee ($P < 0.01$) and lower health expenditures ($P < 0.05$) for Muslims than for Catholics. The fact that Catholics spend more on health may be explained by the effect of extension programs and advice from Catholic Health Centres run by the Catholic Church. In fact, members of the Catholic Church are more informed on, and aware of, health problems. Moreover, the effect of religion on health expenditures could be due to the spread of Catholic churches in the AP area. In rural Senegal, where the Catholic Church plays a prominent part in the provision of health facilities, its members generally have relatively good access to health care (Jutting, 2001; Johannes *et al.*, 2004; Diop *et al.*, 1996; Bitran, 1995). Despite the fact that Catholics generally are better educated than Muslims (30% of Catholics are illiterate compared to 64% of Muslims), we did not find any correlation between education and health expenditures. However, the relation between health status and educational attainment is consistent across diverse racial and ethnic groups (Walton *et al.*, 2005). Nowadays, education is the best indicator of socioeconomic position used in health studies because of the association between health and education found in many empirical studies (Elo and Preston, 1996; Ross and Wu, 1995; Groot *et al.*, 2006).

Men's education is significantly associated with expenditures related to schooling ($P < 0.05$). One percent increase in husband's education leads to 0.91% increase in school expenditures. In fact, husbands' education allows for more attention to children's long-term future and may be connected to their better schooling.

Concluding this section, we can say that the income-pooling hypothesis is partly rejected in this study. Our results also show that the effect of income on expenditures is gender biased with a divergence between husband and wife in expenditure patterns regarding some household goods (health and clothing). However, food provision depends on both the husband's and the wife's income, with the former being more important. Apart from the individual income, education, ethnicity, religion, wife's savings and area, matter in determining gender expenditure patterns.

5.3 Separate utility functions in household decision-making

The rejection of income pooling for all goods and the evidence of the existence of separate-sphere contributions to expenditures for husband and wife in rural households in Senegal, show the weakness of the unitary model of the household and the rejection of common preferences of all members. In our previous findings, earnings are not pooled and are not used jointly by husband and wife in a common household expenditure plan. Instead wife and husband tend to contribute to expenditures separately. In West-African households, as our results have already shown, the male as the 'household head' receives much of the income generated in common activities, even if all members participate in providing labour (wives, children). However, in personal activities, individual members are decision makers for their own earnings (for example, the wife's earnings from milk production and from her own food-crop plots). Based on this argument, our choice for the analysis of utility in rural Senegalese households is the non-cooperative model. Moreover, there is neither evidence from the literature nor from our own findings, of communication between the two partners with respect to the management and the way of spending their individual income. Each partner tries to contribute to the household well-being in an individual way taking the contribution of the other as given. Also, as we have seen in Chapter 3, husband and wife are specialized in different household tasks and according to Smith and Chavas (1999), regarding the specialization of partners in productive and reproductive activities, from the point of view of the household the non-cooperative approach can be applied. This specialization is embedded in gender roles in labour supply and expenditures, where females play a specific role. The analysis of the decision-making process has also highlighted the monetary compensation of income between husband and wife. Other studies have shown that this transfer flows generally from male to female as a compensation for the female's labour supply in common fields especially in cash crop production (Venema, 1986; Guyer, 1988; David, 1991; Dey, 1997; Lilja *et al.,* 1996). In our findings, we observed that this monetary compensation (or transfer) from male to female applies in both areas (see Chapters 3 and 4). The understanding of such a transfer suggests that the decision-making process should take account of preference heterogeneity and spouses' interdependence. Each spouse maximizes his or her well-being, taking the other partner's behaviour as given. This results in an equilibrium which implies an allocation that maximizes the individual utility of each partner. Lastly, our choice of a non-cooperative model also takes into account critiques from other studies, including the misperception of individual interests with respect to their own appreciation of welfare (Sen, 1990). For example, perhaps women's well-being is not primarily related to their own interest, but rather to the welfare of their children or husbands or to that of the entire household.

In the following sections, we develop a non-cooperative model to capture the value that each partner attributes to private goods versus household goods, and we estimate these values.

5.3.1 Specification of the non-cooperative model

Household decision-making theory will be applied to a household with two members: husband and wife, each of whom has a utility function that depends on his or her consumption of private goods and household goods: u^h for the husband and u^f for the wife. The non-cooperative model implies that each partner provides household goods voluntarily; making choices that are maximize utility, given the action of the other partner. In this model, each spouse receives a benefit from joint consumption. The model assumes that the non-pooling outcome is a result of a gender specialization in the provision of household goods where only one partner makes a positive contribution. Good that generates utility for both partners and for children can be considered household good.

In a household represented by husband (h) and wife (f), in decision-making, each member allocates his or her income (y^i) ($i = h,f$), to spending on the household public goods, x_1, and private or personal goods, x_2. We assume that each member's utility depends upon the aggregate level of consumption of household goods ($x_1 = x_1^h + x_1^f$) and personal goods (x_2^i). In our study, household goods are related to food, health and schooling. Private goods are clothing and other unknown goods (x_2^i):

$$y^i = (px_1^i + px_2^i),\ i = h,f \tag{5.7}$$

1 is a household good with price p_1 (both husband and wife face the same price), 2 is a private good with price p_2^h for the husband h, and for the wife f a private good with price p_2^f. The price p_1 is normalized to one for the household goods since partners face the same price of goods, and may face different prices for private goods.

In case of a voluntary transfer between husband and wife, for example a transfer θ from h to f, we have $y^f + \theta$ and $y^h - \theta$ respectively, as the disposable income for female and male. Because for each type of expenditure (private or not) one spouse's contribution to household goods enters into the other's welfare function, husband's or wife's expenditures depend on the way in which each of them expects the other to spend. Thus the interdependence in the household operates only through the consumption of household goods. This leads to Cournot equilibrium where each spouse maximizes his or her well-being, taking the other's behaviour as given. The equilibrium solution of x_1 and x_2 are those that maximize the attitude of each partner towards consumption of private goods. The utility maximization of the problem subject to the budget constraint (1) results in the optimal solutions for husband (h) and wife (f). Theta turns out to represent 7.90% of women's income and 1.5% of husband's income appears to be a relatively minor factor in household decision-making. For this reason, and also because we cannot estimate it jointly with the equilibrium values of the parameters attached to the valuation of the private goods, we treat it as exogenous in our model.

Given x_1^h and x_1^f and given ordinal utility functions u^f, u^h: $R_+^2 \rightarrow R$, positive (prices) p_1, p_2^h, p_2^f, non-negative (budgets), y^f, y^h and $\theta \in R$ such that $y^f + \theta \geq 0$ and $y^h - \theta \geq 0$, we can consider the following two maximization problems:

$$MAX_{x_1^f, x_2^f}\ u^f(x_1^f + x_1^h, x_2^f) \tag{5.8}$$

under the restriction

$$p_1 x_1^f + p_2^f x_2^f \leq y^f + \theta \tag{5.9}$$

$$MAX_{x_1^h, x_2^h}\ u^h(x_1^h + x_1^f, x_2^h) \tag{5.10}$$

under the restriction

$$p_1 x_1^h + p_2^h x_2^h \leq y^h - \theta \tag{5.11}$$

Let $\bar{x}_1^f$ be the Marshall demand function of f for household good 1. This function depends on the constants p_1, p_2^f, $(y^f + \theta$, but more importantly on x_1^h. So let us write $\bar{x}_1^f(x_1^h)$ for this maximizer. Let $\bar{x}_1^h$ be the Marshall demand function of h for household good. Of course this function depends on the constants p_1, p_2^h, $(y^h - \theta$, but more importantly on x_1^f .So let us write $\bar{x}_1^h(x_1^h)$ for this maximizer. The problem we are interested in is to solve values for x_1^f and x_1^h such that:

$\bar{x}_1^f(x_1^h) = x_1^f$, $\bar{x}_1^h(x_1^f) = x_1^h$ (equilibrium conditions).

Our first problem concerns the existence and uniqueness of a household equilibrium. In order to analyze this problem we make the following simplifying assumptions:

1. u^f and u^h are locally non-satiated.
2. For all parameter values p_1, p_2^f, y^f and θ, the wife's maximization problem has a unique solution. This solution depends on the parameters, but most importantly on x_1^h. So let us write just $((\bar{x}_1^f(x_1^h), \bar{x}_2^f(x_1^h))$ for this maximizer. For all parameter values p_1, p_2^h, y^h *and* θ, the husband's maximization problem also has a unique solution. The latter depends on the parameters but most importantly on x_1^f. So let us write just $((\bar{x}_1^h(x_1^f), \bar{x}_2^h(x_1^f))$ for this maximizer.
3. The previous assumption in 2 defined functions such that:
   $\bar{x}_1^f$: $R_+ \rightarrow R_+$ and $\bar{x}_1^h$: $R_+ \rightarrow R_+$ are continuous. It also implies that:
   $\bar{x}_2^f$: $R_+ \rightarrow R_+$ and $\bar{x}_2^h$: $R_+ \rightarrow R_+$ are continuous.

The consequence of assumption 1 is that:

$$p_1 \bar{x}_1^f (x_1^h) + p_2^f \bar{x}_2^f (x_1^h) = y^f + \theta \qquad \text{and} \tag{5.12}$$

$$p_1 \bar{x}_1^h (x_1^f) + p_2^h \bar{x}_2^h (x_1^f) = y^h - \theta \tag{5.13}$$

implying that:

$$\bar{x}_1^f \in \left[0, \frac{y^f + \theta}{p_1}\right], \; \bar{x}_1^h \in \left[0, \frac{y^h - \theta}{p_1}\right].$$

Finally, it follows that $((x_1^f, x_2^f)), (x_1^h), (x_2^h)) \in R_+^2 \times R_+^2$ is a household equilibrium if and only if $\bar{x}_1^f (x_1^h) = \bar{x}_1^f, \bar{x}_1^h (x_1^f) = \bar{x}_1^h$.

Now consider a general Cobb-Douglas utility function:

$U(x_1, \ldots x_n) := x_1^{\alpha_1} \ldots\ldots x_n^{\alpha_n}$ where the $\alpha_i > 0$

$x_1, x_2, \ldots. x_n$ is the quantity of goods consumed.

The maximization of the utility functions leads to an expression for the Marshall demand functions $\bar{x}_i (p_1, p_2 \ldots, p_n; y)$ of the utility function:

$$U = (x_1 + c_1)^{\alpha 1} x_2^{\alpha 2 \ldots} \cdots x_n^{\alpha n} \tag{5.14}$$

where $c_1 > 0$ (c_1 is a constant and a general representation of (x_1^i)). To this end we consider the system of equations:

$$\frac{p_1}{p_j} = \frac{\alpha^1}{\alpha_j} \frac{x_j}{x_1 + c_1} (j = 2, \ldots., n), \tag{5.15}$$

(i.e. the price ratio equals the marginal rate of substitution), and

$p_1 x_1 + p_2 \ldots + p_n x_n = y$ (budget constraint).

Re-writing results in:

$$p_j x_j = \frac{\alpha_j}{\alpha_1} p_1 (x_1 + c_1)..(j = 2 \ldots., n) \tag{5.16}$$

So,

$$p_1 x_1 + p_1 (x_1 + c_1) A = y \text{ where } A = \sum_{j=2}^{n} \frac{\alpha_j}{\alpha_n} \tag{5.17}$$

With,

$$x_1 = \frac{y - p_1 c_1 A}{p_1(1+A)}, \tag{5.18}$$

Now, we have:

$$x_j = \frac{\alpha_j}{\alpha_1}\frac{p_1}{p_j}\left(\frac{y - p_1 c_1 A}{p_1(1+A)}\right) + c_1 (j = 2,...n). \tag{5.19}$$

If we consider two Cobb-Douglas utility functions with:

$$u^f(x_1, x_2) := x_1^{\alpha 1^1} x_2^{\alpha 2^f}, \text{ utility of the wife} \tag{5.20}$$

$$u^h(x_1, x_2) := x_1^{\alpha 1^1} x_2^{\alpha 2^h}, \text{ utility of the husband.} \tag{5.21}$$

Because the utility function is strictly quasi-concave and differentiable on the interior of its domain with merely positive partial derivatives, the solutions found are correct as long as none of them is negative. (In case of negativity there is a boundary solution). Using Equations 5.17 and 5.18 with $n = 2$, the equilibrium conditions yield:

for the household expenditures,

$$\bar{x}_1^f(x_1^h) = \frac{y^f + \theta - p_1 x_1^h \alpha_2^f}{p_1(1+\alpha_2^f)}, \tag{5.22}$$

$$\bar{x}_1^h(x_1^f) = \frac{y^h - \theta - p_1 x_1^f \alpha_2^h}{p_1(1+\alpha_2^h)} \tag{5.23}$$

and for the private expenditures,

$$\bar{x}_2^f = \alpha_2^f \frac{p_1}{p_2^f}\left(\frac{y^f + \theta - p_1 x_1^f \alpha_2^f}{p_1(1+\alpha_2^f)} + x_1^h\right); \text{ and} \tag{5.24}$$

$$\bar{x}_2^h = \alpha_2^h \frac{p_1}{p_2^h}\left(\frac{y^h - \theta - p_1 x_1^f \alpha_2^h}{p_1(1+\alpha_2^h)} + x_1^f\right) \tag{5.25}$$

These equilibriums are the conditional equilibrium values depending on theta (θ) and the price of private goods (p_2^h *and* p_2^f) since husband and wife face the same price of household goods (p) where each individual's equilibrium behaviour is the best response to the equilibrium of the other partner. This leads to a non-cooperative equilibrium in the provision of household goods namely food, health, schooling and private expenditures (clothing and unknown private goods). At the equilibrium condition, the values of x_1^f and x_1^h (5.22) and (5.23) depend upon the price of private goods, the value of transfer, the husband's and wife's incomes and the parameters α_2^f (5.24) and α_2^h (5.25) which measure the value that husband and wife accord to private goods or in economic terms, the elasticity of the utility over the consumption of private goods. For econometric purposes, α_2^f and α_2^h can be solved in terms of x_1^f, x_1^h, the amounts that husbands and wives allot to expenditures on household goods:

Equations 5.26 and 5.27 or in terms of x_2^f *and* x_2^h on private goods: Equations 5.28 and 5.29:

$$\alpha_2^h = \frac{y^h - \theta - x_1^h p_1}{p_1 \left(x_1^h + x_1^f\right)} \tag{5.26}$$

$$\alpha_2^f = \frac{y^f + \theta - x_1^f p_1}{p_1 \left(x_1^h + x_1^f\right)} \tag{5.27}$$

$$\alpha_2^h = \frac{p_2^h x_2^h}{p_1 \left(x_1^h + x_1^f\right)} \tag{5.28}$$

$$\alpha_2^f = \frac{p_2^f x_2^f}{p_1 \left(x_1^h + x_1^f\right)} \tag{5.29}$$

We expect that with higher transfers (θ), the wife will show the greatest α for private goods, because of the high variability of expenditures on household goods with her income (Equation 5.24). In the estimation of α, the income used in the analysis was the income restricted to expenditures for household goods and private goods as reported in the questionnaires. Finally, because utility is ordinal, we may assume that the parameters of the Cobb-Douglas functions can be expressed in a different way such that their sum equals one:

$$\alpha *_2^f = \frac{\alpha_2^f}{\alpha_2^f + 1} \text{ and, } \alpha *_2^h = \frac{\alpha_2^h}{\alpha_2^h + 1} \tag{5.30}$$

5.3.3 Results

The parameters considered in Equations 5.26-5.29 take into account two sources of income: (1) the total labour income comprising the income from agricultural and off-farm activities and (2) the income from cash transfers from migrants which is considered non-labour income. In fact, in Senegal, the labour market outside the household is very small. For this reason, a greater opportunity to get involved in off-farm activities implies greater bargaining power for a spouse (Manser and Brown, 1980). Also, empirical studies brought out evidence that control over non-labour income affects behaviour (Schultz, 1990; Duncan, 1990; Lundberg and Pollak, 1993; Wards-Batts, 2003). According to Pollak (2005), more than earnings (labour income), non-labour income is a relevant argument in the estimation ofbargaining power. In our study, we found that husband and wife separately receive non-labour income (cash transfers from migrants). The share of cash transfers in total income may be another relevant indicator of bargaining power in the context of rural livelihood in Senegal.

The results in Table 5.6 show the average estimates of the private goods parameters in Equation 5.30. The estimation of α in Equations 5.28 and 5.29 using 5.30 can be called utility elasticities, because they measure the responsiveness of utility to a change in the level of expenditures of either private or household goods. The equilibrium values show heterogeneity in preferences between husband and wife: $\alpha_2^h > \alpha_2^f$ and $0 \leq \alpha_2^i < 1$ $(i = h, f)$. Table 5.6 shows estimation results of α for husband and wife, and across areas. Since $\alpha_1 + \alpha_2 = 1$, the results show that wives attach more value to household goods than men. The elasticities for private goods are 0.26 for the wife and 0.33 for the husband. The results suggest that a 1% increase in men's consumption of private goods leads to a 0.33 increase of utility. The wife's expenditure of household goods remains important for her utility and even increases with additional money transfers to her from her husband ((sensitivity to utility over consumption of household expenditures (health, schooling, food)). This finding suggests that with additional income, rural Senegalese women tend to attach more utility to the consumption of household expenditures (schooling, health, food) and less to personal goods (clothes). These results are in line with findings in the UK (Lundberg *et al.*, 1997) arguing that child tax allowance paid to the mother resulted in a significant increase in consumption of children's clothing. Results from Canadian data (Browning *et al.*, 1994) showed that the wife's share of income increases with her share of private expenditures. However, in the case of poor rural farms, the household expenditures remain a priority in the daily consumption of goods and might increase (more than expenditures on clothes) when the wife gets additional income from her husband. In the SP area, with additional transfers, women as well as men attach substantial value to private goods (the wife's elasticity is 0.34 compared to 0.35 for men). This finding may be explained by the fact that SP women are economically more comfortable and thus attach more value to private goods than AP women (whose sense of utility from the consumption of private goods is only 0.18). The equilibrium elasticities in the two situations show that in the case of more bargaining power, i.e. more income, women also accord importance to the expenditure of household goods. This result strengthens our previous findings that schooling, health and food expenditures are elastic with the wife's income. This

Table 5.6. Estimation of α for private goods with voluntary transfers from husband to wife.

	Mean	Std
Entire sample	$\alpha_2^h = 0.33$	0.19
N=299	$\alpha_2^f = 0.26$	0.19
Sylvo-pastoral area	$\alpha_2^h = 0.35^a$	0.17
N=149	$\alpha_2^f = 0.34^b$	0.18
Agro-pastoral area	$\alpha_2^h = 0.31^a$	0.22
N=150	$\alpha_2^f = 0.18^b$	0.16

[a,b] Different superscripts indicate significant differences between areas at $P<0.01$

finding is more visible in the AP area where women earn less (Chapter 2) and where pooling is dominant (Chapter 3). Besides, AP husbands accord more importance to household goods than SP husbands.

Calculations of incomes after transfers from husband to wife indicate that men's income still exceeds that of females (Table 5.7). The transfers in our study represent on average 1.5% of the yearly husband's income. In the theoretical literature of modeling household decision-making (Chen and Woolley, 1999), an increase of transfers would have a positive effect on consumption until these would be spent on female private goods, resulting in decreasing male utility. The income from transfers thus decreases as the wife's income rises, but increases with the husband's income. A small transfer from husband to wife does not affect the husband's welfare, because she will use the entire transfer to purchase household goods. But under fairly general assumptions, an increase in the value of theta from husband to wife will indirectly increase the consumption of women's household expenditures, and correspondingly decrease the male's consumption of private goods. According to empirical tests (Lundberg, 2005), the value of transfer depends on the relative bargaining power of the two spouses and is determined through a bargaining process constrained by social, economic and cultural factors. For example, asset ownership, education, property rights or income opportunities outside marriage may give the wife more bargaining power regarding the value of the transfer. We expect that women who own few assets in the household and are restricted in terms of mobility and decision-making, have little bargaining power and may not be able to influence the value of the transfer in order to increase their consumption of goods. Moreover, motherhood's responsibilities may reduce women's involvement and investment in the labour market and thus reduce her relative bargaining position in relation to men (Baker *et al.*, 2005).

In our empirical results (Table 5.8), the determinants of the amount transferred from husband to wife are the wife's time input in livestock, decision-making in finance management and food provision, and area (AP or SP). As the wife's time allocated to livestock increases, the transfers from husband to wife decrease. Indeed, activities

Table 5.7. Descriptive statistics of the value of monthly income without and with voluntary compensation from husband to wife.

Income (cash transfers from migrants not included)	Mean	Std
Total husband's income before voluntary transfer	143,236	153,280
Total husband's income after voluntary transfer	117,018	112,247
Total wife's income before voluntary	25,447	25,154
Total wife's income after voluntary transfer	30,388	65,255

Table 5.8. Determinants of the transfers from husband to wives.

Variables in log	Coefficient	t-value
Level of education (W) [2]	0.670	0.90
Level of education (H) [2]	-0.396	-0.76
Time allocated to livestock (W)	-0.316	-2.63**
Time allocated to crops (W)	0.0255	0.26
Time allocated to livestock (H)	0.0252	0.44
Time allocated to crops (H)	-0.0241	-0.24
Wife's savings	3.14×10^{-6}	1.39
Wife's childcare	2.939	1.13
Wife's access to land [3]	-2.122	-1.43
Wife's access to credit [4]	-0.247	-0.17
Wife's health status [8]	1.420	0.54
Polygamy [6]	0.148	0.35
Religion [5]	-4.427	-1.19
Household size	-0.0039	-0.03
Management of finance [7]	2.266	3.65***
Responsibility for borrowing food [7]	-1.652	-1.94*
Responsibility for borrowing money to buy food [7]	0.047	0.15
Wife's mobility	0.133	0.198
Dummy Wolof [9]	5.457	2.42**
Dummy Peulh [10]	4.242	1.69
Zone [1]	8.9	4.25***
Constant	24.06	2.07
Adjusted R^2	0.26	

*$P<0.05$, **$P<0.01$, ***$P<0.001$.

W = wife, H = husband.

[1] Dummy (0 = SP area, 1 = AP area).

[2] Dummy (0 = lowest, 1 = highest).

[3] Dummy (1 = yes).

[4] Dummy (1 = yes).

[5] Dummy (0 = Catholic, 1 = Muslim).

[6] Dummy (1 = polygamous household).

[7] (1 = husband only, 2 = husband more than wife, 3 = equally, 4 = wife more than husband, 5 = wife only).

[8] Dummy: 0 = illness problem, 1 = good health.

[9]Dummy Wolof (Wolof = 1, Sereer = 0).

[10] DummyPeulh (Peulh = 1, Sereer = 0).

related to livestock, especially milk production and marketing, constitute an important component (77%) of wife's income from farming. This finding as expected shows that the wife's earnings reduce the relative share of income transferred from her husband, suggesting that women earning more income benefited less from their husbands in terms of cash transfers. In fact, transfers from husbands to wives are a sort of compensation for their labour input in farming (Guyer, 1981; Dey, 1997; Jones, 1983). Depending on the type of society, this compensation depends on the relative bargaining power of women. Because of their poor prospects outside marriage (World Bank, 2001), in many African societies women have little bargaining power within marriage. In our study, earnings from livestock, in particular milk production and ownership of small ruminants, directly benefit women. This situation induces fewer transfers from husband to wife but reflects more women'power (see also Doss *et al.*, 2008) compared to that of wives working in cash crops, whose earnings are entirely controlled by men (Darity, 1995; Lim Sung So *et al.*, 2007. From our results, in response to an increase of the wife's time allocated to livestock, the husband reduces his compensation to her ($P < 0.01$). In most African societies, women have no access to returns from cash cropping and are often oriented towards other income-generating resources (Dey, 1997; Kabeer, 1992; Smith, 1999). The effect of wives' earnings from livestock may constitute more bargaining power to wives within marriage, even at the expense of a smaller husband's income share and thus less compensation.

Empirical studies in South-Ethiopia show that women's bargaining power has a negative impact on coffee production when bargaining power is measured by the size of the woman's livestock herd in case of divorce (Lim Sung So *et al.*, 2007).

Our results also show that an increase in the wife's responsibility for food provision does not result in more compensation from the husband; in fact there is even a negative effect on the transfer. This finding suggests that even if women participate in food provision (by borrowing or buying food), the value of transfers would not increase accordingly, suggesting that the crucial role of women in daily food provision does not affect the value of compensation. This finding shows that men remain the only managers of most of the income and act as those principally responsible for the value of the transfer. The only factors having a positive association with the value of the compensation are the living area and the wife's involvement in the management of finance. Women's responsibility for finance management induces more bargaining power in the transfer, may be through men's confidence in women. Regarding area and ethnicity, being Wolof increases the amount transferred, beyond its already higher average level in their (AP) area. This finding strengthens our conclusion concerning the individualistic behaviour of agents in the SP area compared to those in the AP area, and confirms the more patriarchal social structures of Peulh and Sereer. The computation of the beta coefficients shows substantial values for the wife's time allocated to livestock, her involvement in management of household finance and area (respectively 0.25, 0.21 and 0.32). These variables are more important in explaining the compensation from husband to wife.

5.3.4 Impact of external factors on the equilibrium values of the α parameter

The values of the preference parameters may be related to some important exogenous factors whose analysis is essential. For example, customary gender laws concerning access to common property (i.e. land) are important determinants of the non-cooperative equilibrium (Pollak, 2005; Sevilla-Sanz, 2005; McElroy, 1990). We assume that in the context of rural Senegalese households, the values of the parameters may express a voluntary contribution equilibrium within the marriage that might be affected by socio-cultural factors. In fact, because of religious power (Muslims) and customary practices, divorce is forbidden in most communities and occurs only in special circumstances caused by violence or physical incapability of one the spouses. According to cultural norms, the wife's long-term gain from marriage would be a better future for her children. For example, Islam as the dominant religion in the country, imposes legislation and regulation, on thought and knowledge, thus embedding the inferior position of women (Mbow, 2009). Besides, more than religion, social norms cause the gender gap in decision power (Dowd *et al.,* 2009). When assessing the bargaining relationships between husband and wife, in light of Sen's criticisms (Sen, 1990) consideration should be given to the effect of social power within communities. For example, the equilibrium values reflect the bargaining process because they show the vulnerability or the strength of the persons involved at a certain time. In the context of our empirical study, we try to capture this effect through the relative status of the female's autonomy in decisions regarding food security and mobility. According to Sen (1990), the members of the household face two problems simultaneously. One involves cooperation, the other has to do with conflict. Also, the perceived interest response can be seen as the value that a person attaches to his or her own welfare compared to the perceived welfare of the other. The appreciation of the breadwinner orientations of both female and male for especially food-security purposes, as well as polygyny can be seen as indices of perceived interest.

Finally we consider the most important factors that may affect the perceived contribution to the equilibrium outcome of the bargaining game. In the context of our study these factors may be related to the level of education because of the high level of illiteracy of both husband and wife, also to their health status which may affect their farm income.

The regression analysis on α_2^* for women's private goods, has an adjusted R-squared of 0.035 (Table 5.9); for men, the adjusted R-squared is only 0.09, suggesting that in both models the independent variables considered do not adequately explain

Table 5.9. Determinant of the α^* parameters (for private goods).

Variables in log	Determinants of α_2^f		Determinants of α_2^h	
	Coefficient	t-value	Coefficient	t-value
Constant	-0.964	-0.159	-9.168	-0.649
Religion [4]	2.758	1.307	6.469	1.212
Zone [1]	-1.328	-1.375	-1.091	-0.482
Polygamous [5]	-0.023	-0.178	-0.125	-0.200
Husband's education [2]	0.087	0.234	0.633	0.904
Wife's education [2]	0.040	0.226	0.977	0.952
Dummy Wolof [7]	-1.748	-1.715	-4.661	-1.86*
Dummy Peulh [8]	-2.896	-2.31**	-3.715	-1.228
Wife's access to land [3]	-0.817	0.980	-0.703	-0.376
Husband's health status [6]	0.226	0.224	1.0119	0.288
Wife's health status [6]	0.115	1.307	1.542	0.035
N	298		266	
R^2 adjusted	0.035		0.09	

*$P<0.05$, ** $P<0.01$.
W = women, H = husband.
[1] Dummy (1 = AP area).
[2] Dummy (0 = lowest, 1 = highest).
[3] Dummy (1 = yes).
[4] Dummy(0 = catholic, 2 = muslum).
[5] Dummy (1 = polygamous household).
[6] Dummy: 0 = illness problem, 1 = good health.
[7] Dummy Wolof (Wolof = 1, Sereer = 0).
[8] Dummy Peulh (Peulh = 1, Sereer = 0).

the equilibrium values. Unlike ethnicity, which explains the parameters in the two models, the other variables (wife's access to land, education, polygamy) seem not to be associated with the husband's and wife's valuation of the household's private expenditures. While being Wolof increases the husband's parameter, being Peulh is negatively associated with α_2^f.

5.4 Conclusion

In this chapter we investigated how husband and wife in rural households participate and behave towards consumption expenditures. The results show that there are differences in preferences since the income-pooling hypothesis is rejected in this study. However, there is a predominance of pooling in the AP area. The hypothesis that one additional franc of wives' income is spent the same way as one additional franc of husbands' income is rejected for expenditures related to health, clothing, tea and coffee. For food and schooling, it seems that the partners could spend income in a consensual way, respecting a sort of sharing between husband and wife. Also, our results show that the effect of income on expenditures is gender biased: one additional franc for men is more likely to be spent on clothing and health than an additional franc for women; and for women an additional franc is likely to raise schooling expenditures. Food provision depends on both the husband's and the wife's income with the former being more important. Area also appeared to matter: the wife's income significantly increased food, health and schooling expenditures in the SP, but not in the AP area. For AP men, an additional franc significantly increased clothing expenditures.

The investigation of expenditure determinants showed that consumption also depends upon individual and community characteristics. While the results concerning household size show that an additional member of the household raises expenditures for almost all goods except tea and coffee, area and ethnicity show a difference in priorities given to consumption. Indeed, in the AP area where households earn less, the consumption of clothes, tea and coffee significantly decreases for each additional franc earned. Moreover, regarding ethnicity, among the Peulh one additional franc is less likely spent on health and clothing, and more likely on food and schooling. Among the Wolof, expenditures on tea, coffee and clothes were significantly lower than for the reference (Sereer). Being a catholic is positively associated with expenditures related to health, and negatively associated with those related to tea and coffee. Their relatively higher expenditures on health can be related to the Catholic Church they belong to, which emphasizes more extension programs mostly related to health care and sickness prevention. Finally, better educated male heads show a higher propensity to spend on schooling, thus leading to positive effects on child education and school attainment.

The non-cooperative model shows that taking into account money compensation from husband to wife, men value private goods more than women, resulting in a higher share in the consumption of their private expenditures. However, overall, the values of the parameters show that both husband and wife are more sensitive to the consumption of household goods (food, schooling, health) than to the consumption of private goods (clothing).

Transfers from husband to wife lead to a higher valuation of her consumption of household goods in the equilibrium. These transfers increase when we move from SP area to AP area. Also, the wife's responsibility in the management of finance is positively associated with the transfers she gains from her husband. However, in the SP area, because of their relatively comfortable economic position, women tend to value private goods equally high as men do. The transfer from husband to wife seems to be more important in the AP area where pooling of income is dominant. Looking into the exogenous variables that may be associated with the non-cooperative equilibrium shows that as expected, ethnicity (expressing difference in community behaviour) is correlated with the equilibrium values of husband's and wife's valuation of private goods.

Chapter 6

Intrahousehold resource allocation: analysis of subjective welfare and well-being

6.1 Introduction

Chapter 6 extends the analysis of individual behaviour by using subjective measures of welfare and well-being. This approach of the Leyden School differs from the neo-classical methods because it uses direct questioning. The chapter provides empirical evidence of different attitudes between genders towards the valuation of income (in terms of monetary satisfaction) in rural Senegalese households. Furthermore, it extends the concept of income satisfaction to the estimation of poverty lines derived from the subjective welfare approach. Finally, the chapter considers the wider concept of well-being including life, health and work satisfaction.

6.1.1 Theory and concept

Utility has been historically assumed to capture feelings of pleasure or pain (Bentham, 1789, reprinted in 1970). According to Kapteyn (1985), utility is assumed to increase if enjoyable activities are undertaken. 'Welfare economics' as a notion emerged in the 1900s to analyze effects of policies and economic activities on the welfare of the population. The old 'welfare economics' refers to major assumptions strongly linked to cardinal utility (having cardinal qualities), and interpersonal comparisons of utility). This approach has later been criticized by economists who in the late 1930s developed the 'New Welfare Economics' based on the principle of Pareto efficiency and on the concept of perfectly competitive markets. The measurement of utility was challenged because utility values of alternative goods could not be measured on the basis of observed static consumer choice (Robbins, 1932). For example, if cardinal measurement would be possible, it would be reasonable to say '*I prefer oranges to apples twice as much as I prefer rice to cereals.*' This example summarizes the possibility of being able to compare preferences on the same scale, e.g. the amount by which the utility from eating oranges exceeds that from eating apples. Accordingly, regarding interpersonal comparisons, it would be possible to say that '*I prefer bananas to oranges more than you do.*' By challenging this possibility, economists finally restricted the welfare concept to ordinal utility and to the concept of Pareto efficiency. However, Pareto efficiency, meaning a policy change with no losers, implies many different allocation possibilities and the impossibility to make the right choice. In practice, economists have continued to be dissatisfied with the restricted ordinal concept (Ng, 1997). In fact decision-making dealing with different time periods (savings, investment decisions) needs a higher than ordinal measurement level. The problem of utility as a measure of welfare was described by Wansbeek and Kapteyn (1983: 249) who said *'Utility seems to be to economists what the Lord is to theologians. Economists talk about utility all the time, but do not seem to have hope of ever observing it this side of heaven. In micro-economic theory, almost every model is built on utility functions of some kind. In empirical work little attempt is made to measure this pervasive concept'.*

There is a growing interest in the use of cardinal utility which concerned three approaches: (1) measurement of a utility function also assuming its functional form; this approach was used in our previous chapter to measure the utility of husband and wife with respect to expenditures; (2) an axiomatic approach to utility, in which it is not measured but only assumed (Christensen *et al.*, 1975; Jorgenson *et al.*, 1984); and (3) measurement of utility using both subjective and objective indicators. Economists concerned with the latter approach are interested in examining the living conditions of poor people and their 'basic needs' through poverty measurements (Ravallion, 1994, Sen, 1987). Also, in the latter approach, certain economists examined the quality of life of nations (Nussbaum *et al.*, 1992; Sen, 1987; Maasoumi *et al.*, 1989) including a range of variables (literacy level, infant mortality rate, etc.), the quantification of which many economists considered controversial. Finally, economists are concerned with the quality of individuals' lives, which can be linked to the work of Clark *et al., 1996*, on 'unhappiness' implying an indirect measure of experienced utility (Van Praag *et al.*, 1999). Other economists were interested in the individual's evaluation of economic outcomes by performing probability choice experiments on individuals. This approach was criticized because individuals are not good at dealing with probabilities, as they overestimate small probabilities and underestimate large ones (Allais and Hagen, 1979). Another approach concerning subjective welfare and well-being, assumed that individuals were able to describe their utility levels by means of verbal qualifications. Our approach follows the subjective measurement of welfare (Van Praag, 1968, 1971; Van Praag and Ferrer-i-Carbonell, 2004) and underpins our analysis of behaviour described in Chapter 5. The subjective welfare approach originated at Leyden University in the Netherlands in the early 1970s. Van Praag postulated a theory to operationalize the concept of experienced utility which allows a direct cardinal measure of utility. This theory implies the possibility for individuals to evaluate their income position on a finite interval scale located between the 'worst possible situation' and the 'best possible situation' (Plug, 1997). The resulting Welfare Function of Income (WFI) is an instrument for associating welfare levels with different levels of income on a [0,1] scale.

6.1.2 Review of relevant findings of the Leyden Welfare approach

Referring to empirical findings concerning subjective welfare, the Leyden approach focuses primarily on the evaluation of income, showing that welfare functions differ between individuals. These differences can be correlated with observable individual or environmental characteristics (Van Praag *et al.*, 1999). A study on poverty by Van Praag, Hagenaars, and Van Weeren in eight European countries (1982) and another in

Russia (Frijters and Van Praag, 1995), showed that variables such as household size, current and past income, and age, affected the individual welfare function. Results from dynamic estimations showed that current income had the largest weight and income in the past influenced the individual welfare function much more than future income (Van Praag and Van Weeren, 1983). Relevant empirical policy applications are related (1) to the evaluation of income redistribution (definition of a subjective poverty line for individuals) and tax policy, (2) to the determination of the level of adaptation of individuals facing changing socio-economic situations and (3) the evaluation of the impact of inflation on income norms and on the satisfaction with individual income. The Welfare function of Income (WFI) is relevant in estimating individual welfare levels in different situations, hence the effect of different situations and policies can be compared. Another direct question referring to the minimum income a respondent feels he needs in his circumstances to make ends meet is relevant to studying subjective poverty (Goedhart *et al.,* 1977; Calosanto *et al.*, 1983; Kapteyn *et al.*, 1985). Van Praag *et al.* (1982a, 1982b) used a second approach based on the translation of the Welfare Function of Income (WFI) into a minimum income level in member countries of the European Community. Elsewhere, the WFI is used by Van Praag (1977) and Kapteyn (1977) to estimate income inequality between individuals; in fact, income inequality is perceived differently by different individuals, and policy interventions that can be perceived as egalitarian by some, may be seen by others as increasing inequality (Kapteyn and Wansbeek, 1985). More recently, the Leyden welfare approach applied to the household has extended the estimation of the WFI to the happiness of people living together, i.e. husband and wife. Using a German dataset, the welfare functions of both spouses and their interdependence were studied (Plug, 1997). The subjective welfare approach is aimed at understanding three concepts. First, the Individual Welfare Function can be estimated using the income evaluation question (IEQ); based on an individual's assessment of incomes. Second, welfare can be estimated using the Minimum Income Question (MIQ) related to the minimum income that individuals report as sufficient to make ends meet. And thirdly, the WFI and the MIQ can be used in order to estimate the Subjective Poverty Line (SPL) and the Leyden Poverty Line (LPL) to be described in Section 6.2.

Economic welfare referring to the evaluation of income should be distinguished from well-being, which refers to the evaluation of other important notions of life, such as health, family and friends, marriage and work satisfaction, physical environment and political satisfaction, etc. In many psychological and sociological surveys, people were able to evaluate their life as a whole (see Cantril, 1965; Veenhoven, 1996; Diener and Suh, 1997). Problems concerning health, work, the relationship with the partner, sleep or neighbourhood significantly affected well-being but hardly affected welfare (Plug and Van Praag, 1995b). German studies suggest that happiness is influenced by gender, marital situation (Diener *et al.*, 1999), and health status (Myres, 1999; Van Praag *et al.,* 2000). Similar findings are reported for job satisfaction, religion and education (Clark and Oswald, 1994; Diener *et al.*, 1999; Ellison, 1991). Coming to the correlation between welfare (subjective income evaluation) and subjective well-being, Van Praag *et al.* (2000) found that financial satisfaction, more than the absolute level of income, is associated with subjective well-being. This finding corroborates the fact that the perception of income named 'subjective evaluation of income' (as the extent to which an individual is satisfied with his own income in order to cover needs), has a higher correlation with subjective well-being than with the level of income (see Schyns, 2001; Dimoso, 2009).

The Leyden approach is relevant in the context of our study. Indeed, since the concept of welfare has usually been related to income (Samuelson, 1956; Becker, 1974; Singh *et al.*, 1986; Ravallion, 1994), subjective evaluation of income might take different meanings in the context of developing countries. For example, gender is a key determinant of food security, i.e. current and future daily nutrition of household members (Niehof, 2003). Because of their caretaking in the food system and daily household needs, women may evaluate the same resources and assets differently than men, so their welfare perceptions may differ. Different welfare perceptions may result from different individual needs, in particular needs related to gender. For example, a person with high-level needs will perceive the welfare associated with a particular income as lower than a person with low-level needs.

While a wide body of literature has analyzed welfare through the estimation of a utility function, we would like to provide more insight by using the Leyden approach, assuming that individuals are able to describe directly the welfare associated with their income, by means of verbal qualifications (Van Praag, 1971; Van Praag *et al.*, 1995). Since welfare is usually limited to economic issues, such as income, we will also study the concept of well-being, capturing the individual evaluation of a number of life domains e.g. health, family, work and education. In order to study the wider impact of gender roles, health status and labour, this study will assess how people judge their own well-being (Antonides *et al.*, 1980, Diener and Suh, 1997; Cornelisse *et al.*, 2006). Moreover, the Leyden approach has been extended to the analysis of poverty situations using two subjective methods: the Leyden Poverty Line and the Subjective Poverty Line. These poverty lines describe a specific relationship between subjective welfare and income. They are identified through subjective methods using the stated preferences of the persons studied (Plug, 1997; Hagenaars, 1986) instead of previously used methods where policy makers determined who is poor or not. Considering the high level of poverty in these countries, the estimation of poverty lines may be relevant in research programs and projects in other developing countries

So far, the Leyden approach has only been tested in developed countries and not yet in developing countries, except for the recent work of Dimoso (2009) in Tanzania. This chapter tests the hypothesis that partners have different opinions or attitudes towards subjective welfare, towards the relation between household income and minimum needs, and towards subjective well-being. It applies the Leyden approach to husband and wife separately in order to find out whether the partners have different responses towards the relation between household income and minimum needs. The notions of household needs and minimum income raise two important types of questions to be considered in our methodological approach (Goedhart *et al.*, 1977; Danziger *et al.*, 1984; Van Praag and Van der Sar, 1988).

6.2 Methods

The Leyden approach focuses primarily on the assessment of incomes in connection to welfare. The approach deals with 'utility of income', 'income satisfaction' or 'economic welfare.' The measurement approach is based on two assumptions: first, individuals are able to evaluate individual incomes by means of *verbal qualifiers* using 'good,' 'bad,' 'sufficient,' etc., and second, verbal qualifiers can be translated into numerical evaluations on a [0,1] scale. In this study information has been gathered by the following two questions put to both spouses:

1. Taking into account your situation, size of your household, your family and farm and other activities, state your income per month (in CFA) that you would find very bad, bad, sufficient, insufficient, good and very good.

Evaluation	Incomes stated by husband	Incomes stated by wife
Very bad		
Bad		
Sufficient		
Insufficient		
Good		
Very good		

2. What is the minimum household income you need in your situation to make ends meet?
   'In my circumstances, I would need about .... CFA to make ends meet.'

6.2.1 The individual welfare function approach

The first welfare question is the so-called income evaluation question (IEQ) (Van Praag *et al.*, 1999); the second is the minimum income question (MIQ). In order to find a reliable estimate of the relationship between income and utility, one needs to consider between 4 and 9 levels of income (Van Praag, 1971, Plug, 1997); in our study, we use 6 p.6/5levels of evaluation. The IEQ has been estimated successfully by many other researchers in economics, sociology and psychology (Plug, 1997, Kapteyn *et al.*, 1988).

In the first question, if the number of verbal qualifications equals k, the welfare level corresponding to the i-th verbal equals $(2i\text{-}1)/2k$ (Parducci, 1995), assuming equally spaced welfare intervals. For each respondent, we have six income levels connected to six utility levels. Van Praag (1968) assumed that the welfare distribution function $U(y)$ would be a lognormal distribution function which is defined as:

$$\begin{aligned} U(y;\mu,\sigma) &= N(\ln y;\mu,\sigma) \\ &= N((\log y-\mu)/\sigma;0,1) \\ &= \Lambda(y;\mu,\sigma) \end{aligned} \tag{6.1}$$

Where $N(\mu,\sigma)$ stands for the normal distribution function with mean μ and variance σ^2, and $\Lambda(\ln y;\mu,\sigma)$ is the lognormal distribution function with parameters μ and σ. As y refers to income in this study, $U(y)$ is called the individual welfare function of income (WFI). The two parameters μ and σ vary across individuals. The parameter μ is the log income needed to attain 0.5 on the [0,1] utility scale. As the parameter μ increases, the individual becomes less satisfied with the same amount of income and needs more income to reach the same welfare level. The parameter σ indicates the sensitivity to different incomes, defining the slope of the welfare function. A steep slope would indicate sharply increasing utility with an increasing income. The parameters μ and σ are estimated for each individual by:

$$\mu = \frac{1}{k}\sum_{i=1}^{k}\ln(c_i) \tag{6.2}$$

$$\sigma_i = 1/(k-1)\sum(\ln c_j-\mu_i)^2 = \sqrt{\frac{1}{(n-1)}\sum_{n=1}^{6}(\ln c_{in}-\mu_i)}^{\,2} \tag{6.3}$$

where $c_1, \ldots c_n$ stand for the six income levels.

The minimum income question (MIQ) measures income poverty (Goedhart *et al.*, 1977; Plug, 1997; Danziger *et al.*, 1984).

The IEQ and MIQ may depend on other variables such as the size of the household, age, education, time allocation, gender and decision-making of the spouses and current income. The following relations have been investigated:

$$\mu_i = \beta_0 + \beta_1 \ln fs_i + \beta_2 \ln y_{i,c} + f(X_i) + f(Z) + \varepsilon_i \tag{6.4}$$

$$y^i_{min} = \delta_0 + \delta_1 \ln fs_i + \delta_2 \ln y_{i,c} + f(X_i) + f(Z) + \partial_i \tag{6.5}$$

$$f(Z) = f(ag, ed, gd, Ti) \tag{6.6}$$

y_{min} is described as the income level associated with the welfare level 'absolutely minimal.' y^i_{min} denotes individual minimum income, ε_i and ∂_i are random disturbance terms, fs_i (family size) denotes number of individuals living in the household of respondent i, $y_{i,c}$ is current income of individual i, Z denotes a number of variables used in the regression analysis such as age (*ag*), education (*ed*), living area (*are*), time allocation to crops and livestock (*Ti*), and ε_i *and* ϵ_i are error terms.

It is expected that the welfare parameters μ_i and y^i_{min} increase with the number of persons in the household (fs_i) and with current individual income ($y_{i,ci}$). Concerning the former effect, children in the household create costs and therefore impact on the welfare level. The income effect implies that people adapt their income judgments to changes in their current income (Van Praag, 1971). It is also assumed that aspirations increase when income increases, so people adapt their income evaluation to changes in their current financial situation. A higher income usually induces higher needs, leading to less welfare associated with particular incomes. In this study we only consider the effect of current income, keeping in mind that past income or future income may further affect the WFI. However, due to a lack of data on past and anticipated income in our sample, their effect is ignored. Two individuals A and B with two current incomes $y_{A,c}$, $y_{B,c}$ may evaluate their income differently. Thus, the following is usually true:

$$U\,(y_{Bj};\, fs,\, y_B) \neq U\,(y_{Bj}\, fs,\, y_A) \tag{6.7}$$

Also, income levels may differ according to groups of people, so welfare evaluation is determined by the social reference group to which individuals belong. For this reason, we include the living area (*are*) as a relevant explanatory variable of the welfare function. The variable age (*ag*) is relevant because we may expect that at the beginning of our life the WFI increases and may decrease later on. However, recent studies brought out that in the first stage of our life there is a negative correlation between age and welfare until at least the 30s and 40s (Van Praag *et al.*, 2000; Oswald, 1997). Education of respondents (*Ed*) may also impact on the WFI directly or indirectly because of its possible effect on employment (off-farm activities) and on improvement of farming systems (equipment, better seeds, and better technology). Because time allocation in production (T_i) is an input in production- generating income, we also expect that it will be associated with welfare. Other variables related to decision-making and power in the household may also be associated with individual welfare and individual appreciation of the minimum income.

In our study, the questions concerning gender are addressed to husband and wife separately. The IEQ and the MIQ of the husband (*h*) and wife (*w*) using Equations 6.4 and 6.5 are explained by the following relations:

$$\mu^h = \beta_{0h} + \beta_1^h\, ln\, fs + \beta_2^h\, ln\, y_c + \beta_3^h ln\, ag^h + \beta_4\, ln^2\, ag^h + \beta_5^h\, ln\, ed^h + \beta_6\, lnTi + \beta(Z) + \varepsilon^h \tag{6.8}$$

$$\mu^f = \beta_{0f} + \beta_1^f\, ln\, fs + \beta_2^f\, ln\, y_c + \beta_3^f\, ln\, ag^f + \beta_4\, ln^2\, ag^f + \beta_5^f\, ln\, ed^f + \beta_6\, lnTi + \beta\,(Z) + \varepsilon^f \tag{6.9}$$

$$ln\, y^h{}_{hin} = \alpha_{0h} + \alpha_1^h\, ln\, fs + \alpha_2^h\, ln\, y_c + \alpha_3^h ln\, ag^h + \alpha_4\, ln^2\, ag^h + \alpha_5^h\, ln\, ed^h + \alpha_6\, lnTi + \beta(Z) + \eta^h \tag{6.10}$$

$$ln\ y^f_{min} = \alpha_{0f} + \alpha_1^f\ ln\ fs + \alpha_2^f\ ln\ y_c + \alpha_3^f\ ln\ ag^f + \alpha_4\ ln^2\ ag^f + \alpha_5^f\ ln\ ed^f + \alpha_6\ lnTi + \alpha\ (Z) + \eta^f \quad (6.11)$$

Living area and ethnicity group are captured by dummy variables. In fact the appreciation of welfare may differ according to ethnic group because of the subjectivity of the importance that individuals assign to values and tastes. Results from Chapter 3 showed that people in the sylvo-pastoral area have more income than those in the agro-pastoral area. Hence we may expect that the parameter μ increases as income increases and that individuals in the sylvo-pastoral area value their income differently than those in the agro-pastoral area. The parameters β_2 and α_2 express this effect through the notion of 'preference drift' (Van Praag, 1971), meaning that individuals adapt the judgment of their income according to the level of income in their living area. The disaggregation of the IEQ and MIQ by gender allows the analysis of response behaviour of the household and tests for similarities and differences across gender with respect to feelings towards welfare and behaviour.

6.2.2 Welfare and Subjective Poverty

The traditional equation explaining the income evaluation parameter (Van Praag and Plug, 1995) is:

$$\mu_i = \beta_0 + \beta_1 \ln fs + \beta_2 \ln y_{ci} \quad (6.12)$$

where μ increases with respect to y_c and *fs*. The linkage between poverty and income evaluation can be constructed based on several poverty lines; one of these is defined as a certain percentage γ_0 of the average μ_y in society (Plug, 1997). This poverty line y_1 can be expressed as:

$$y_1 = \gamma_0 \int_0^\infty y dF = \gamma_0 \mu_y \text{ where } 0 \leq \gamma_0 \leq 1 \quad (6.12.1)$$

The definition of being poor is represented as having less income than y_1. The poverty percentage can then be expressed as:

$$\Pi_1 = \int_0^{y_1} dF \quad (6.12.2)$$

The poverty concept as defined by Goedhart *et al.* (1977) and Hagenaars (1986) yields a specific relationship between welfare *u* and income y. In this relationship, the welfare of the poor falls under a *critical welfare threshold* μ_χ, and the poverty line can be defined by the relation $u(y_\chi) = \mu_\chi$. The Leyden Poverty Line (LPL) can then be expressed by the equation:

$$\kappa(y_\chi; \mu(y_\chi), \sigma) = \mu_\chi \quad (6.12.3)$$

The LPL yields the equation:

$$\ln y_\chi = \frac{\beta_0 + \beta_1 \ln fs + \chi\sigma}{1-\beta_2} \tag{6.12.4}$$

Since the critical welfare threshold is represented frequently as being about halfway the Leyden welfare interval, the value of χ may be assumed to equal 0.4 or 0.5. The expression of the LPL includes a correction for the shift of the welfare function caused by obtaining the poverty line income instead of their current income (when reporting their income evaluation).

Through the minimum income question (MIQ), Goedhart *et al.* (1977) determined the subjective poverty line (SPL). Individuals who have an income lower than the minimum income level y_{min} cannot make ends meet. So through the MIQ, the poverty line y_{min}, is the solution to the equation:

$$ln\, y_{min} = \alpha_0 + \alpha_1 lnfs + \alpha_2\, ln\, y_{min} \tag{6.13}$$

According to Equation 6.12.4, the subjective poverty line (SPL) is estimated to be:

$$ln\, y_{min} = \frac{\alpha_0 + \alpha_1 \ln fs}{1-\alpha_2} \tag{6.13.1}$$

The SPL represents the correction due to the fact that the welfare function shifts when the minimum income is actually obtained (once individuals receive their minimum income, they are happier with the incomes they stated before and the welfare function shifts to the right). The estimation of the Subjective Poverty Line and the Leyden Poverty Line for the 300 households representing our sample in the two areas will be given in due course.

6.2.3. Subjective well-being

As a wider concept than welfare, subjective well-being encompasses other aspects of life such as health status, work satisfaction, etc. Alternatively, it may be called quality of life. Empirically, it is possible for individuals to evaluate their life as a whole. We use the well-known question developed by Cantril (1965): *'here is a ladder with ten steps which denotes the 'ladder of life.' The bottom step stands for the worst possible life. If you climb up and arrive at the tenth step, you arrive at the best possible life. Can you indicate where you are at the moment?'* For convenience, in our survey question put to husband and wife separately, we use seven steps instead of ten, as follows:

'Here is a picture of a ladder. If the top of the ladder represents the best possible life for you and the bottom represents the worst possible life for you, where on the ladder do you feel you personally stand at the present time?'

7. Best possible life
6
5
4
3
2
1.Worst possible life

This question was addressed with respect to the general situation of husband and wife separately, and also for their health and their work.

Well-being, *W*, depends on a number of explanatory variables (Plug and Van Praag (1995a) and Plug (1997). Some of the variables are related to objective factors such as education, age, family size, income, religion, etc., and others are called 'problem intensities.' These latter variables denote the intensity of experiencing problems with their health, their work, their physical environment, etc. Well-being can be written as: $W = W(z)$ where z denotes a vector of problems experienced and objective variables. In the context of developing countries, as we have already seen in Chapter 4, health problems impact on time allocated to ill persons, resulting in loss of income and households facing additional costs related to traditional or modern treatment and transportation. We might expect that people facing health problems (H) will see their well-being decrease. Land ownership (ld) which determines the quality of crops and livestock production, might also be associated with subjective well-being. The objective variables are a subset of predictors variables included in the welfare model estimation, such as household size (fs), individual income (y_c), age (ag) and age squared, and education (ed). Other variables (H) might also influence the well-being of individuals, like the chance to get involved in off-farm activities for diversification purposes and to intervene in decision-making, especially for women. The well-being function can then be estimated for husband (W^h) and wife (W^f) separately written as:

$$W^h = \beta_0 + \beta_1^h \ln fs + \beta_2^h \ln y_c + \beta_3^h \ln ag^h + \beta_4 \ln^2 ag^h + \beta_5^h \ln ed^h + \beta_6^h \ln H + \beta_7^h \ln ld + \varepsilon^h \tag{6.14}$$

$$W^f = \gamma_0 + \gamma_1^f \ln fs + \gamma_2^f \ln y_c + \gamma_3^f \ln ag^f + \gamma_4 \ln^2 ag^f + \gamma_5^f \ln ed^f + \gamma_6^f \ln H + \gamma_7^f \ln ld + \eta^f \tag{6.15}$$

In each equation, the area of research, religion and ethnicity are considered as dummies. Since well-being is expressed from a ladder with seven levels of possibilities responses, we will estimate our equations using multinomial logistic regressions. We use this method because we assume that the 'distance' between the different point of the ladder are not equal. For example, the distance between *worse possible life* and *bad life* may be shorter than the distance between *good life* and *best possible life'*. Data were taken from our sample of 300 households from different agro-ecological areas and different reference groups with respect to agricultural activities and income levels (see Chapter 3). The IEQ and the MIQ as well as the 'quality of life' questions have been put separately to both husband and wife (who are the major breadwinners of the household).

6.3 Results

6.3.1 Individual welfare through estimation of IEQ and MIQ

Results of estimation

The subjective welfare function of income has been estimated for 300 couples represented by husband and wife separately, using the Income Evaluation Question (IEQ) and the MIQ. The estimation assumes that the Individual Welfare Functions $U(y)$ differ between individuals and is gendered. We recall that μ determines the location of $U(y)$ and denotes the log-income evaluated at 0.5 on the [0,1] scale, and that $U(y,\mu)$ is decreasing in μ.

Descriptive statistics in Table 6.1 show that the average household incomes reported in answering the IEQ vary across zones and are higher for men than for women. According to the results, in evaluating the household income at the same level, males need more income than women. Across zones, estimates also reveal significant differences between male household heads (mhh). In the SP area they estimate higher incomes as sufficient than their counterparts in the AP area. The average value reported as *sufficient* is 185,000 CFA for SP mhh and 117,680 CFA for AP mhh. For women, *sufficient* income is almost the same in the two areas: 76,400 CFA in the SP and 74,100 CFA in the AP area. Between husbands and wives, the difference

Table 6.1. Average income reported for income evaluation question (CFA)[1].

	Total sample N=300		Sylvo-Pastoral area		Agro-Pastoral area	
Evaluation	Husband	Wife	Husband	Wife	Husband	Wife
Very bad income	67,432**	43,414	78,858	43,057	58,130	43,766
Bad income	92,214**	56,528	114,460	55,440	70,266	57,600
Insufficient income	111,660***	75,243	135,100	76,400	88,530	74,100
Sufficient income	151,430***	95,560	185,638	97,700	117,680	93,450
Good income	189,496***	123,670	225,878	126,000	153,600	121,400
Very good income	248,070***	151,275*	283,040	139,460	213,566	163,000

Results of T-tests (difference between areas): ** $P<0.01$, *** $P<0.001$.

[1] 1 US$ = 475 CFA during the year of the survey.

of average values for *sufficient* income is not surprising and might be explained by the principle role of men in providing the means to fulfil household needs (food, health, clothes) as demonstrated in Chapter 4. Differences across areas with respect to husbands reporting *sufficient* income may be explained by the land quality differences between the two areas and the consequent impact on crops and livestock production (production of crops, cost of feeds and other inputs, scarcity of water for animals and people). Indeed, as we saw in Chapters 2 and 3, households in the SP area face a dry physical environment and are less productive in crop growing. Hence SP households are constrained to spend more on inputs in cereals and other foods to compensate for the low productivity of crops. Women's reported incomes were similar across areas except for *very good income* which is higher in the AP area where women report higher values compared to the SP area.

Table 6.2 shows that the estimated value of the 'want parameter' for the location of the welfare function $\mu(hxu)$ for husbands is generally higher than $\mu(wxu)$ for wives: 11.62 for husbands compared to 11.20 for wives, suggesting that with the same income, men are less satisfied than women. The reported monthly household income evaluated as *sufficient* to cover household needs equalled 151,431 CFA and 95,560 CFA for husbands and wives, respectively. Generally the household income is much lower; the average incomes estimated by our surveys are 106,692 CFA for husbands and 22,272 CFA for wives. These results suggest that in general, households need more income to cover their needs at a sufficient level. The average estimated levels of subjective economic welfare (*huy* and *wuy*), evaluated at the minimum income needed to make ends meet are 0.72 and 0.73 for husbands and wives respectively. The subjective economic welfare seems to be high because it's based on the evaluation of their minimum income instead of their current income. The actual income is lower than the minimum they need. The R-square of the regression for the individual welfare function(WFI) averages 0.93 for husbands and 0.94 for

Table 6.2. Average values of the individual economic welfare parameters.

Zone		hxsig	hxmu	R^2h	huy	wxsig	wxmu	R^2w	wuy
Sylvo-pastoral	Mean	0.50	11.81***	0.91	0.76***	0.47	11.16	0.91	0.77***
N=148	Std	0.25	0.43	0.12	0.15	0.23	0.46	0.15	0.14
Agro-pastoral	Mean	0.49	11.44	0.95	0.67	0.51	11.23	0.96	0.70
N=150	Std	0.25	0.424	0.091	0.20	0.24	0.39	0.074	0.17
Total	Mean	0.50	11.62	0.93	0.72	0.49	11.20	0.94	0.73
N=298	Std	0.25	0.46	0.11	0.18	0.23	0.43	0.12	0.16

Results of T-test (difference between areas): *** $P<0.001$

wives, indicating reliable estimates, considering the number of welfare levels used (Van Praag *et al.*, 1999).

In response to the IEQ, male heads in the SP area report higher incomes than those in the AP area, suggesting that for a given level of happiness the former need more money than the latter. Figure 6.1 shows some differences between zones; for women, the charts do not show any significant differences between the two areas. The T-tests show significant differences between zones with respect to the welfare function of men, and the subjective welfare of the total household income for men and for women separately.

Turning to the minimum income question (MIQ), Table 6.3 shows gender differences, suggesting that husbands and wives differ in their appreciation of household

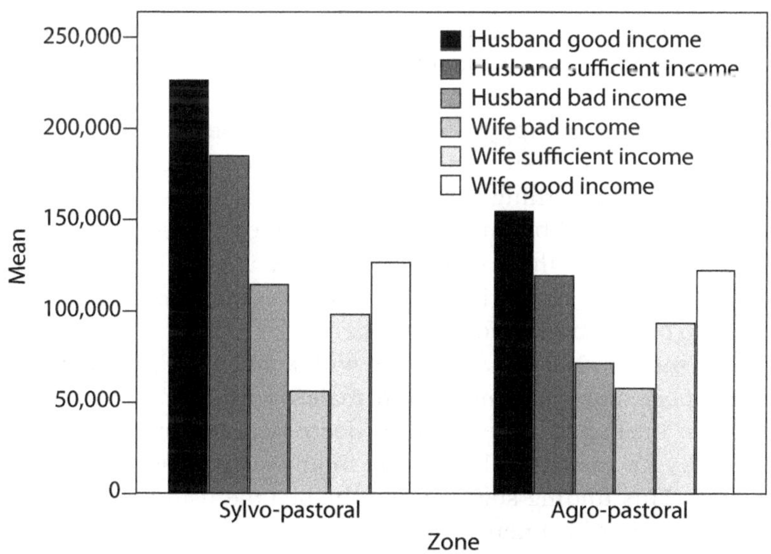

Figure 6.1. Distribution of the incomes reported for the income evaluation question (IEQ) by zone.

Table 6.3. Average incomes reported for the income evaluation question (in CFA).

	Total sample N=300		Sylvo-Pastoral area		Agro-Pastoral area	
	Husband	Wife	Husband	Wife	Husband	Wife
Minimum income	187,212**	122,470*	210,000	113,000	164,664	132,000

Results of T-test (difference between areas): * $P<0.05$, ** $P<0.01$.

expenditures. Husbands reported higher MIQ than wives, suggesting that women need less money to make ends meet than men. This may be explained by men's breadwinner role, as we observed in Chapter 4. These incomes are much higher than the actual incomes, which points at a great welfare gap.

Determinants of subjective welfare through IEQ and MIQ

The results in Table 6.4 of the subjective welfare function parameter μ_i (h_{xmu} *for husband and* w_{xmu} *for wife*) shows the dependence of the individual welfare function for husband and wife as specified in Equations 6.8 and 6.9. For husbands the R^2 equalled 0.22; for women it was 0.23. Males in the SP area appear to be less happy with a given income than those in the AP area. In contrast, wives in the AP area are less happy with a given income than those in the SP area. Since the effect of the minimum income needed is positive, this indicates that when the income increases, both husbands and wives will become less happy with a given income: this effect testifies to the fact that at higher income levels, people are more ambitious and less easily satisfied with their income (preference drift, see Section 6.2.1). Expenditures on clothes make men less happy ($P<0.01$): 1% increase in costs related to clothes leads to a 0.13 increase in the income that the husband needs to reach utility at 0.5 on the welfare scale. As clothing constitutes an important part of household needs that are primarily handled by husbands (see Chapter 5), the increase of this type of expenditures, given the income, may negatively affect his welfare. For women, significant determinants of the individual welfare function parameter are men's time allocated to crops and to decision-making. The less time men allocate to crop production, the more unhappy women are with a given income. This would mean that women expect more earnings when males work on crops and become less happy when the husbands' time in crops decreases. Regarding decision-making, less responsibility for managing household finance makes women less happy with a given income. This finding suggests that involvement in decision-making is an important component of women's welfare in Senegal.

The regression analysis of the MIQ in Table 6.5 shows that from individual characteristics, Equations 6.10 and 6.11 explain households' MIQ quite well, especially for men. For husbands, the average R^2 of the minimum income regression equals 0.28 compared to 0.15 for wives. The reported minimum income is positively associated with the husband's actual monthly income. A 1% increase in monthly income leads to a 0.17% increase in the required minimum income. However, the wives' reported minimum incomes do not vary with their actual incomes. Another determinant of the required minimum income for husbands is expenditures on clothes. The results for wives show that the physical location of the farm influences their required minimum income. Women in the AP area reported higher values of minimum income than women in the SP area, probably because of the prices of goods in this area. Regarding people's ethnicity, being Sereer (living in the AP area) is negatively associated with

Table 6.4. Estimates of the regression analysis of the location of the individual welfare function (μ).

Variables in log	Husband		Wife	
	β coeff.	t-value	β coeff.	t-value
(Constant)	10.74	9.714	9.446	12.710
Age (log)	0.050	0.307	0.083	0.768
Zone [2]	-0.420	-3.325**	0.228	1.990*
Household size (log)	0.117	1.058	0.156	1.877*
Monthly minimum income	0.100	2.105*	0.070	2.514**
Polygamous [3]	0.026	1.215	0.031	1.569
Level of education [1]	-0.004	-0.051	0.093	1.339
Religion [4]	-0.369	-1.365	0.161	0.668
Log time in livestock (H)	-0.023	-0.751	0.040	1.541
Log time in livestock (W)	0.008	0.138	-0.020	-0.481
Log time in crops (H)	-0.052	-1.358	-0.062	-1.980*
Log time in crops (W)	0.053	1.533	0.032	1.024
Who borrows food for the household [5]	-0.002	-0.071	0.012	0.709
Who borrows money [5]	0.004	0.179	0.019	1.177
Who usually buys food [5]	-0.026	-0.415	-0.099	-2.540**
Who manages finance in the household [5]	-0.009	-0.235	-0.112	-3.98***
Expenditure on food in log	-0.048	-0.847	0.001	0.099
Expenditures on clothes in log	0.130**	2.629**	0.012	1.607
Expenditures on health in log	-0.025	-0.570	0.008	1.096
Expenditures on school in log	0.046	1.115	-0.002	-0.236
Dummy Peulh	0.101	1.180	0.123	1.781
Dummy Wolof	-0.021	-0.165	-0.151	-1.375
Number of observations	168		220	
Adjusted R^2	0.22		0.23	

* $P<0.05$, ** $P<0.01$, *** $P<0.001$.
H: husband, W: wife.
[1] Dummy (0 = lowest, 1 = highest).
[2] Dummy (0 = SP area, 1 = AP area).
[3] Dummy (0 = monogamous household, 1 = polygamous household).
[4] Dummy (0 = Catholic, 1 = Muslim).
[5] (1 = husband only, 2 = husband more than wife, 3 = equally, 4 = wife more than husband, 5 = wife only).

Table 6.5. Estimates of the regression analysis of the minimum income.

Variables in log	Husband		Wife	
	B	t-value	B	t-value
(Constant)	9.340	6.296	11.071	10.499
Age in log	0.005	0.021	0.074	0.485
Zone [2]	-0.146	-0.860	0.658	4.060***
Household size in log	0.168	1.130	0.027	0.227
Monthly actual income	0.173	2.630**	-0.003	-0.069
Polygamous [3]	0.048	1.65	0.031	1.114
Level of education [1]	-0.080	-0.794	-0.071	-0.718
Religion [4]	-0.474	-1.300	-0.291	-0.857
Log time in livestock (H)	0.040	0.956	0.029	0.766
Log time in livestock (W	-0.080	-1.096	0.083	1.389
Log time in crops (H)	-0.035	-0.686	0.001	0.021
Log time in crops (W)	0.069	1.501	-0.029	-0.651
Who borrows food for the household [5]	-0.010	-0.331	-0.006	-0.240
Who borrows money [5]	0.014	0.514	-0.005	-0.225
Who usually buys food [5]	-0.076	-0.907	-0.103	-1.860*
Who manages finance in the household [5]	-0.013	-0.233	-0.105	-2.620**
Expenditure on food	-0.062	-0.824	0.012	1.145
Expenditures on clothes	0.198	2.84**	0.018	1.643
Expenditures on health	0.085	1.443	0.012	1.149
Expenditures on school	0.003	0.052	-0.012	-0.895
Dummy Peulh [7]	0.127	1.105	0.123	1.264
Dummy Wolof [6]	-0.537	-3.080**	-0.429	-2.74**
N	166		219	
Adjusted R^2	00.28		0.16	

* $P<0.05$, ** $P<0.01$, *** $P<0.001$.

H: husband, W: wife.

[1] Dummy (1 = highest).

[2] Dummy (0 = SP area, 1 = AP area).

[3] Dummy (0 = monogamous household, 1 = polygamous household).

[4] Dummy (0 = Catholic, 1 = Muslim).

[5] (1 = husband only, 2 = husband more than wife, 3 = equally, 4 = wife more than husband, 5 = wife only) dummies ethnicity.

[6] (1 = Wolof, 0 = Sereer).

[7] (1 = Peulh, 0 = Sereer).

the reported minimum income. The women's responsibility in managing finance is negatively associated with reported minimum income. The estimated results suggest that the less women are involved in managing finance, the higher the minimum income needed. For expenditures, no significant effect was observed, suggesting that women appreciate the costs of household needs differently than men and make the same amount more valuable in terms of expenditures. The results finally suggest that partners in the same household do not fully agree on the income needed to make minimum household ends meet. This finding may relate to the differences in earnings between husband and wife who primarily focus on their earnings when answering the MIQ and their financial situation. The gender difference in answering the MIQ is also found in studies conducted in some developed countries (Plug and Van Praag, 1995a; Plug, 1997).

6.3.2 Leyden poverty line estimates

The estimation of Equations 6.13 and 6.14 are shown in Table 6.6. The results of the regressions show significant correlations between household income and welfare. The adjusted R^2s are particularly high for the estimation of the minimum income, 0.75 and 0.68 respectively for husband and wife, suggesting that the welfare parameters are well explained by household size and household income. Women's welfare is significantly associated with household size; an increase in household size induces more needs for women in the household. However, for men welfare is more significantly associated with household income than for women. Estimates from the minimum income also show that men's welfare is slightly more elastic with respect to household income than women's welfare.

Table 6.7 describes the poverty lines, using the different methods of estimation (LPL and SPL). The estimates show significant differences between the LPL and the SPL for each gender. The SPL shows much lower values than the LPL. This result is in

Table 6.6. Estimates of the regression of μ^i and y^i_{min}.

	μ^h		μ^f		$\ln y^h_{min}$		$\ln y^f_{min}$	
	Mean	Std	Mean	Std	Mean	Std	Mean	Std
Intercept	10.16	0.37	9.70	0.36	-0.64	0.44	0.64	0.43
Ln household size	0.03	0.08	0.28	0.07***	-0.06	0.05	0.06	0.05
Ln household income	0.14	0.03***	0.08	0.03**	1.08	0.036***	0.91	0.04***
Adjusted R^2	0.07		0.09		0.75		0.68	

***$P < 0.001$, **$P < 0.01$.

agreement with Goedhart *et al.* (1977) who found that the SPL was evaluated at about 0.3. Obviously, LPL04 and LPL05, corresponding with welfare levels of 0.4 and 0.5, respectively, are much higher than the SPL. With respect to earnings in the two areas of our sample and the level of poverty in rural Senegal (67% of farmers are poor, according to objective UN evaluation methods: IMF, 2006), the SPL values seem to be more realistic. The SPL and LPL show a significant difference between genders (Table 6.7). The SPLs are 11,384 CFA for women and 16,155 CFA for men, and as for the LPL approach, estimates show that the values of male poverty lines are much higher than those of female poverty lines. With respect to the SPL measure, 58.72% of women in the sample are poor compared to 44.63% of men. According to Van Praag and Van der Sar (1988), these findings suggest that men have higher needs than women. The SPL and the LPL also show significant differences between the two areas. With respect to the LPL and the SPL, the AP area shows higher poverty lines than the SP area, suggesting that AP farmers feel poorer than SP farmers. This finding may be related to the costs of household needs which should be much higher in the AP area.

In all cases, the poverty line estimates for women have the lowest values suggesting that husbands and wives with similar household characteristics do not feel poor at the same level. According to the subjective method, the results suggest that men feel poorer than women with the same income. For example, with respect to individuals who have off-farm jobs (Table 6.8), the poverty lines indicate that in comparison to women, men require a 36% increase in income in order to maintain the same welfare level. The same can be said for education: in comparison to better-educated women, better-educated men need a 43% increase in income to be equally happy. Moreover,

Table 6.7. Poverty lines estimates (CFA)[1].

	Total sample	SP area	AP area
$LPL04_f$	94,845***	90,219	99,700
$LPL05_f$	98,715***	94,000	103,700
$LPL04_m$	176,310***	176,300	178,000
$LPL05_m$	185,349***	185,349	187,200
SPL_f	11,384***	10,000	12,700
SPL_m	16,155***	14,185	18,000

T-test (difference between areas and between genders): ***P<0.001.

[1] 1 US$ = 475 CFA during the year of the survey.

Table 6.8. Poverty line estimates by socio-economic factors (in CFA)[1].

Factors	Subjective poverty line (female)	Subjective poverty line (male)
Husband with lowest education	10,938	15,522
Husband with highest education	12,457	17,677
Wife with lowest education	11,384	16,155
Wife with highest education	11,499	16,318
No off-farm jobs	10,938	15,522
Availability of off-farms jobs	13,767	19,732
Wife without access to micro-credit	11,159	15,835
Wife with access to micro-credit	11,968	16,984

[1] 1 US$ = 475 CFA during the year of the survey

within the same gender group, people with access to off-farm job opportunities feel poorer with the same amount of income than those without such access.

6.3.3 Well-being under the subjective Leyden approach

Regarding well-being, results of the answers given in Table 6.9 show that the majority of respondents (both men and women) reported the mid-level of well-being regarding their overall life, their health and work. Regarding their health and work situation, male heads are happier in the SP area than in the AP area. More women than men

Table 6.9. Well-being by gender and well-being by domain (percentage).

	Husband			Wife		
	Life	Health*	Work***	Life	Health	Work
1 Worst possible	5.35	3,34	8.03	3.68	4.35	12.04
2	23.74	15.72	33.72	31.44	19.40	40.80
3	47.49	37.46	35.79	41.81	41.14	39.46
4	18.73	30.43	19.73	20.40	30.10	6.02
5	4.35	12.71	3.34	2.68	5.02	1.67
6	0.66	0.33	0.33	0.00	0.00	0.00
7 Best possible	0.00	0.00	0.00	0.00	0.00	0.00

Chi-square tests (difference between genders): $^{*}P<0.05$, $^{***}P<0.001$

perceived their life, health and work as bad or very bad. The chi-squared test shows that for health and work there is a significant difference between genders. With regard to having 'good health' or 'good work', women feel unhappier than male heads.' This finding reflects the general health situation of women who are more affected by health problems and dissatisfaction with income from work.

Looking into the multinominal probit estimation, the likelihood ratio chi-square of 62.70 for husband and 82.33 for wife, with a *P*-value of 0.0002 for husband and 0.000 for wife tells us that our model as a whole is statistically significant. Individuals are likely to be more unhappy in the AP area than in the SP area for both men and women. The results of the estimation of Equations 6.14 and 6.15 show difference in factors predicting subjective well-being for husband and wife (Table 6.10). Subjective well-being seems to be also influenced by the decision-making proxies, income and household size. For husbands, income predicts well-being: one unit increase in husband's income will lead to 0.11 increase in the probability at being happy. Wife's mobility and responsibility in food provision and management of finance are good predictors of husband's well-being. For one unit increase in providing food and managing finance (going from 0 to 1), we expect respectively a 0.33 increase in the probability of being more happy and a 0.39 decrease of the probability of being more happy, given all other variables held constant. Household size predicts well well-being since one unit increase of the size of the household, we expect a 0.73 increase in husband's well-being given all other variables held constant. For wives, we would say that Wolof and Peulh are more unhappy than Sereer. Education and decision-making proxies seem to matter in predicting women's well-being. While education and management of household finance predict an increase of women's well-being, asking permission predicts unhappiness. For wives, the influence of income on well-being is not significant, suggesting that spouses have different ideas on the meaning of well-being. Other factors than income might be more important in predicting the female well-being. When we displayed the results as proportional odds ratios (see Table 6.11), we would interpret the results as from a binary logistic regressions. For example, for zone, we say that for moving from SP area to AP area by one unit (going from 0 to 1 scale), the odds of high well-being (level 6 or 7) versus the combined middle and low levels (for example 3 or 4) are 0.16 greater, given that all of the other variables in the model are held constant. Unlike the result for men, the lack of influence of income on women's well-being suggests that people living in the same household do not appreciate their well-being status in the same way (Plug, 1997). Some objective variables (wife's access to credit, polygamy, religion) do not appear to predict the level of reported well-being. The empirical social literature (Sen, 1987; Veenhoven, 1991) supports the evidence that people have different ideas on the meaning of a *good life,* which may be influenced by values, tastes and preferences that people attach to things and individuals.

Table 6.10. Estimates of the logistic models for husband's and wife's well-being.

Variables in log	Husband			Wife	
	β coeff.	t-value		β coeff.	t-value
Pseudo R^2	0.09			0.12	
Zone [2]	-1.028	-2.02*		-1.773	-3.63***
Age squared (W)	-0.000	-0.13		-0.001	-1.85*
Age squared (H)	0.000	0.67		0.001	1.09
Age of husband	-0.008	-0.95		-0.074	-0.79
Age of wife	0.1835	0.71		0.144	1.67
Land cultivated	0.0203	0.23		0.054	0.62
Husband's income	0.1148	1.88*		0.046	0.61
Wife's income	0.0332	0.90		-0.010	-0.28
Health status of husband [8]	0.986	1.61		0.332	0.60
Health status of wife [8]	-0.482	-0.75		0.139	0.24
Wife's own cattle	0.1354	0.80		0.122	0.74
Wife's own sheep and goats	0.0433	0.31		-0.135	-0.95
Household size	0.734	1.83*		0.628	1.71
Religion [4]	0.0767	0.09		0.476	0.58
Dummy Wolof [9]	-0.956	-1.77		-2.326	-4.51***
Dummy Peulh [11]	-0.735	-1.33		-1.942	-3.65***
Polygamous [3]	0.130	1.47		0.046	0.53
Off-farm Jobs [6]	-0.013	-0.43		0.026	0.85
Education husband [1]	0.195	0.67		-0.109	-0.37
Education wife [1]	-0.181	-0.55		0.677	2.05*
Wife access to micro-credit [7]	-0.205	-0.63		0.432	1.32
Wife 's savings	0.0385	1.47		-0.013	-0.51
Permission factor	-0.333	-2.08*		-0.339	-2.22*
Sharing factor	0.1763	1.12		0.133	0.84
Responsibility in food provision [5]	0.3311	2.39**		0.377	2.75**
Responsibility in management of the household [5]	-0.397	-2.64**		0.1017	0.72
/cut1 [10]	-0.820		/cut1	-2.459	
/cut2	1.512		/cut2	0.686	
/cut3	3.953		/cut3	3.064	
/cut4	5.88		/cut4	5.930	
/cut5	8.40				
N=245			N=254		
LR $chi^2(28)=62.70$			LR $chi^2(26)=82.33$		
Prob $chi^2<0.001$			Prob $chi^2<0.0001$		
Log likelihood=-295.358			Log likelihood=-276.476		

Table 6.10. Continued.

* $P<0.05$, ** $P<0.01$, *** $P< 0.001$.
[1] Dummy (1 = highest).
[2] Dummy (1 = AP area).
[3] Dummy (1 = polygamous household).
[4] Dummy (0 = Catholic, 1 = Muslim).
[5] (1 = husband only, 2 = husband more than wife, 3 = equally, 4 = wife more than husband, 5 = wife only).
[6] Dummy (0 = no, yes = 1).
[7] Dummy (0 = no, 1 = yes).
[8] Dummy (0 = illness problem, 1 = good health).
[9] Dummy Wolof (Wolof = 1, Sereer = 0).
[10] The number of cuts is explained by Table 6.9 (husbands have zero response for level 7 and women have zero response for level 6 and 7).
[11] Dummy peulh (Peulh = 1, Sereer = 0).

Table 6.11. Estimates of the odds ratio for husband's and wife's well-being.

Variables in log	Husband		Wife	
	Odds ratio	t-value	Odds ratio	t-value
Pseudo R^2	0.09		0.12	
Zone [2]	0.468	-1.55	0.169	-3.63***
Age squared (W)	0.999	-0.70	0.998	-1.85*
Age squared (H)	1.00	1.39	1.001	1.09
Age of husband	0.886	-1.28	0.928	-0.79
Age of wife	0.105	0.65	1.155	1.67
Land cultivated	1.018	0.21	1.056	0.62
Husband's income	0.119	1.85*	1.047	0.61
Wife's income	1.035	0.96	0.989	-0.28
Health status of husband [8]	2.577	1.64	1.393	0.60
Health status of wife [8]	0.929	-0.12	1.149	0.24
Wife's own cattle	1.128	0.76	1.129	0.74
Wife's own sheep and goats	1.074	0.51	0.872	-0.95
Household size	1.561	1.20	1.875	1.71
Religion [4]	0.733	-0.39	1.610	0.58
Dummy Wolof [9]	0.451	-1.51	0.097	-4.51***
Dummy Peulh [10]	0.610	-0.91	0.143	-3.65***
Polygamous [3]	1.125	1.34	1.047	0.53
Off-farm Jobs [7]	0.975	-0.80	1.02	0.85
Education husband [1]	1.231	0.73	0.896	-0.37

Table 6.11. Continued.

Variables in log	Husband			Wife	
	Odds ratio	t-value		Odds ratio	t-value
Education wife [1]	0.872	-0.43		1.96	2.05*
Wife access to micro-credit [6]	0.864	-0.46		1.541	1.32
Wife's savings	1.032	1.25		0.986	-0.51
Permission factor	0.699	-2.29*		0.711	-2.22*
Sharing factor	1.186	1.12		1.142	0.84
Responsibility in food provision [5]	1.392	2.43**		1.458	2.75**
Responsibility in management of the household [5]	0.705	-2.44**		1.107	0.72
/cut1 [10]	-2.670		/cut1	-2.459	
/cut2	-0.333		/cut2	0.686	
/cut3	2.155		/cut3	3.064	
/cut4	4.10		/cut4	5.930	
/cut5	6.656				
LR chi^2(26)=62.99					
Prob chi^2<0.001					
Log likelihood=-305.02					

* $P<0.05$, ** $P<0.01$, *** $P<0.001$.

[1]. Dummy (1 = highest).

[2] Dummy (1 = AP area).

[3] Dummy (1 = polygamous household).

[4] Dummy (0 = Catholic, 1 = Muslim).

[5] (1 = husband only, 2 = husband more than wife, 3 = equally, 4 = wife more than husband, 5 = wife only).

[6] Dummy (0 = no, yes = 1).

[7] Dummy (0 = no, 1 = yes).

[8] Dummy: 0 = illness problem, 1 = good health.

[9] Dummy Wolof (Wolof = 1, Sereer = 0).

[10] The number of cuts is explained by Table 6.9 (husbands have zero response for level 7 and women have zero response for level 6 and 7).

[11] Dummy peulh (Peulh = 1, Sereer = 0).

6.4 Conclusion

In Chapter 6 we have investigated welfare, well-being and poverty lines of husband and wife according to the Leyden approach. Furthermore, the study has extended the appreciation of welfare by individuals, using verbal qualifications in the estimation of the poverty lines in rural areas. The estimation of the individual welfare function (WFI) shows significant differences between genders and across areas. The average 'want parameter' of the welfare function was 11.62 for husbands and 11.20 for wives. These values denote the log-income evaluated at 0.5 on the [0,1] welfare scale. The corresponding average incomes in CFA were estimated at 111,300 and 73,130 for husband and wife, respectively. Overall, people need more revenues to make ends meet. The WFI differs between individuals and results brought out that with the same level of income, farmers in the sylvo-pastoral area are less happy with their income than those in the agro-pastoral area. The gender distribution of the income evaluation question (IEQ) shows that men are not happy with their income; for most farmers, the income they reported as a *bad income* is even above their monthly real income. In addition, decision-making in the household (in providing food and managing finance), human capital parameters, time allocation on crops, and expenditures on private goods adequately explained the location of the individual welfare function, indicated by μ.

The Leyden approach has also been successfully used to estimate the Leyden Poverty Line and the Subjective Poverty Line in the two areas. Again, the subjective appreciation of poverty by individuals has brought out that responses differ by gender and ethnicity. The findings suggest that men have higher needs than women, and feel poorer than women with the same income. Furthermore, the subjective poverty lines show higher values for better-educated people, for those with access to off-farm jobs and for wives with access to credit. However, a significant effect is only found for wives' access to credit: women's access to credit induces higher poverty lines.

The extended concept of well-being has incorporated other important notions in our life, such as health status, family, and work satisfaction. Results from the Cantril question suggest that husbands and wives have different ideas on what happiness refers to when it comes to the level of income, but share a priori similar feelings with respect to women's involvement in decision-making. The effect of ethnicity is more visible for women: Wolof and Peulh are more unhappy than Sereer. Other important human-capital variables like education, age and household size are good predictors of happiness.

Chapter 7

Conclusion, discussion and outlook

7.1 Introduction and research question

In most developing countries, agriculture is the pillar of the economy with an important share in GDP. However, the deficiencies in agricultural production are a serious cause of poverty in developing countries. Moreover, the lack of consideration of gender issues has greatly contributed to inefficiencies in agricultural activities (Grown *et al.*, 2000, Hands 2001, World Bank, 1997). In Senegal, the efficiency of farming performance (combining livestock and crop activities) depends on the gender division of labour and resource allocation within the household (land, human capital, time). In Senegal, production types vary across sylvo-pastoral (mainly livestock) and agro-pastoral production systems (the latter combining livestock with cash crops and food crops), the availability of land being one of the critical factors. The intrahousehold dynamics regarding resource allocation constitutes an important determinant of production efficiency and the access to income and welfare of household members. Nevertheless, household members encounter high variability in their resources that determine their well-being. The intrahousehold distribution of tasks, responsibilities and decision-making is gendered because of the specialization of agents, husband and wife in particular. In many communities women are specialized in homework which is time-consuming, but are also very much engaged in performing agricultural tasks on common or individual fields. Consequently, the productivity of agricultural activities depends on a gender allocation of resources.

The knowledge of the determinants of production and income in rural households in Senegal is very limited. For this purpose, we expected that outcome estimated through total earnings from agricultural activities is gender-differentiated. The redivision of income related to the earnings of women and men has an impact on their decision-making and responsibility regarding household public expenditures like health or schooling. It is expected that gender differentiation in earnings and assets is an important factor in women's power and voice in the household that affects its responsiveness to food insecurity and to shocks like health problems of household members. For instance, the fact that women are physically more vulnerable to health problems than men also makes them more vulnerable to the loss of control over resources and economic benefits. For this reason, it is relevant to analyze the extent to which more empowerment of women in terms of earnings and control over income, may improve their health status and that of their children. This concluding chapter attempts to answer the following research questions:

- How does the distribution of intrahousehold activities and of responsibilities in farming differ across sylvo-pastoral and agro-pastoral areas?

- What are the structural and socio-economic determinants that are associated with the gender division of resource allocation and control over income across different systems of production?
- How do the gender allocation of labour and time devoted the agricultural activities, housework and health within the household correlate with the earnings of agricultural activities?
- How do men and women engage in food and health provision in terms of decision-making and how can we measure women's power?
- How can we capture the bargaining relationship between husband and wife and its effects on expenditures, and how can we measure their welfare and well-being?

In the first part of this chapter, we summarize the main findings regarding the gender allocation of resources and the determinants of total earnings related to agricultural production. Furthermore, it pays attention to the results with respect to differences between men and women regarding priorities of expenditures and the extent to which bargaining relationships between husband and wife partly determine the capability of household members to provide food. Since the behaviour of men and that of women differ and, because of the fact that husband and wife have heterogeneous preferences, it is expected that spouses do not place the same value on their personal well-being, the well-being of their children or the well-being of the household as a whole. Thus, husband and wife act differently according to their earnings, their education and some factors related to the socio-cultural environment. Actually, factors influencing decision-making towards resource allocation and expenditures may be a threat to the household's welfare and poverty level. Even though several studies have analyzed the important role of intrahousehold resource allocation in agricultural production, few have captured the gender aspect in the consumption of rural households, in their coping with health problems and food insecurity and the consequences for their well-being. The remainder of this concluding chapter provides the relevance of the findings and makes some recommendations for future research.

The results presented are based on an empirical analysis of the behaviour of rural married couples. Cross-sectional data have been collected in the 2006-2007 period by means of a survey. A total of 600 husbands and wives from 300 couples were questioned on the different aspects mentioned above. The data were gathered from two rural areas in Senegal, the Sylvo-pastoral (SP) and the Agro-pastoral (AP) area. These two areas differ in their socio-economic and agro-ecologic aspects. While the dry SP area is mainly based on livestock, the AP area is a mixed farming system combining livestock and crops. Farmers in the SP area are first and foremost herders and pastoralists. In the AP area, land is a critical constraint that induces competition between keeping livestock and crop growing.

7.2 Methodological aspects of the research

7.2.1 Study design

The study design of the thesis is based on a comparative analysis between two areas with a sample of 150 households in each. The deliberate distinction between two important areas in Senegal takes into account the geographical, social, economic and cultural differences between the two communities, which could be expected to influence the dependent variables. Such a study has never before been conducted in Senegal. Data collection was done by personal interviews at the level of the household, using formal questionnaires. The surveys were conducted by ourselves and with some of the civil servants of the Livestock Board Office at Kaolack and UPPRAL at Dahra in order to overcome difficulties related to reliability in household surveys. To accomplish the task in the survey period, six enumerators assisted in the application of the questionnaires. The enumerators have a long experience in the application of questionnaires and each of them is working close to the farmers, and was living in the rural area concerned. The questionnaires have been applied separately to husband and wife in order to gain insight into the gender differences in earnings, labour time allocation and decision-making towards expenditures. In addition, husband and wife have separately answered the subjective welfare and well-being questionnaires.

In our thesis we use both neoclassical and behavioural economics in the analysis of the differences between husband and wife in decision-making, consumption and subjective welfare and well-being. Several methods of estimation were used in the research.

7.2.2 Analysis of farm production and income

We assumed that there is a gender difference in time allocation. For this reason, we performed disaggregate equations to estimate the determinants of income from crops and livestock. But since there are no income observations for some 9.7% of the sample for livestock and 49% for crops, and to avoid selection bias, we estimated the equations by a more general model that allows participation of partners in agricultural production and agricultural output to be determined by different variables. Moreover, in the estimation of the individual performance of partners in crops and livestock, it has been noticed that individual income is a censored variable. The reason for this is that 14.66% of the males in the sample have zero observations of livestock income, and 40.16% have zero observations of crop income; and that for the females in the sample, 40% censored observations are noticed for livestock and 91% for crops. The model uses Heckman's two-step model to estimate the effects of determinants of the censored incomes (Heckman, 1979). Chapter 3 shows that this procedure yielded sensible results. The probit estimates showed the effect of the explanatory variables on the household's participation in livestock and crop growing and the OLS made use of other explanatory variables to explain the variation of total earnings and the time allocation. The advantage of Heckman's model was that it allowed us to estimate the determinants of participation of all observations in productive activities separately from the determinants of the outcome (earnings or time allocation) for the non-censored (i.e. positive) observations. From this method, we were able to appreciate productivity by estimating the relationship between total earnings and labour and time allocation by gender in crop and livestock.

7.2.3 Game-theoretical model versus unitary approach in the explanation of household behaviour

A non-cooperative model with two utility functions was used in the analysis of household behaviour. The significant difference in estimates for men and women regarding decision-making concerning household expenditures, provides evidence that household income is not pooled. This result is strengthened by the results of the Engel curves for major household expenditures (food, health, schooling and clothing) that show that income is not pooled for food and schooling. This finding contradicts the assumption of a single utility function maximized in the household. Most empirical studies in this research field used an aggregated utility function for the household (Singh *et al.,* 1986; Browning *et al.,* 1994; Udry, 1996; Quisumbing and Maluccio, 2000). In our research we tested a bargaining model with two players (husband and wife) where each player maximizes his or her utility, taking the action of the other player as given. This model has been advocated by several authors in the analysis of household behaviour (Himmelweit *et al.,* 2001, Lundberg and Pollak, 1993, Carter and Katz, 1997). The results gave strong support to reject the unitary model, and provided empirical evidence that husband and wife behave differently in rural households in Senegal. Although the utility functions used in the model include the prices of private goods, the utility parameters were derived from the expenditures of husband and wife on household and private goods. In fact, it would not have been relevant to estimate the parameters using the price data, because of its limited cross-sectional variation (one year of survey). Nevertheless, the lack of panel data partly limited our control of individual heterogeneity and variability over time.

7.2.4 The Leyden approach in the analysis of subjective welfare and well-being

The analysis of welfare and well-being of husband and wife used the Leyden approach (Kapteyn and Wansbeek, 1985; Van Praag and Frijters, 1999; Van Praag and Ferrer-i-Carbonell, 2004). The method employs individual evaluations of income by using direct questioning. In the early stage of the Leyden approach in economics (Kapteyn *et al.,* 1988) the direct questions were related to the income needs of the household and were put to the main breadwinner only. However, in our research we have information from two respondents, namely husband and wife. Our findings support the reliability of the WFI method. The well-being measures derived from the WFI can be explained by social and economic factors.

Furthermore, our results provided additional empirical support for the relation between income and welfare. In contrast, the correlation between well-being and income has been found insignificant in most studies and was higher among poor countries compared to richer countries. Our findings conclude that for women subjective well-being is not related to income, thus showing that objective variables did not fully explain individual subjective well-being (Diener *et al.*, 1993). The inclusion of social and subjective variables in the estimation such as ethnicity, decision-making, religion and polygamy, provided substantial empirical evidence that *internal factors* (Diener and Lucas, 1999) are very important in determining well-being. Our study limited the factors to decision-making proxies and socio-economic determinants, but excluded non-economic factors like feelings of misery and other emotions that may affect people, especially the poor (Schwartz and Strack, 1991; Diener *et al.*, 1991). Such inclusions may be captured in future research.

7.3 Summary of the findings concerning the relevant hypotheses

This part of the chapter summarizes the findings of our research in four sections: (1) the first one is related to the variability of land use and the diversity of agricultural activities, (2) the second section gives insights into the labour and time used, and the relationship between total earnings and socio-economic factors that may affect agricultural activities, (3) the third section discusses decision-making concerning food provision and expenditures, (4) the fourth section analyzes the bargaining relationships between husband and wife. The last section provides our concluding remarks with respect to the Leyden approach. In the different parts mentioned above, we tested the following hypotheses:

1. The first and second sections test the hypothesis that women's time allocation to farming is negatively affected by housework and verify the hypothesis that household and individual's earnings depends on individual, household and socio-cultural characteristics.
2. The third section tests the hypothesis that decision-making towards household food provision and time allocated to health care and treatment is gendered, and the status of women in the household in terms of decision-making is positively associated to their bargaining power.
3. In the fourth section, we test the hypothesis that a priori, income is not pooled in the household in the two areas, and that husband and wife engage in a sort of cooperation in order to reach the well-being and the food security of the household.
4. Finally, the fifth and final section tests the hypothesis that partners have different opinions or attitudes towards subjective welfare and the relation between household income and minimum needs, and towards subjective well-being.

7.3.1 Variability in land use and diversity of agricultural activities

The study took place in two different rural areas in Senegal, the Sylvo-pastoral area (SP) and the Agro-pastoral area (AP), which differ in asset endowments and socio-cultural environments. The two areas contrast because of the variability in agro-ecological activities, support programs by government and NGOs in micro-credit organizations, extension programs and training, and other facilities (hospitals, health centres, and access to markets). While the Sylvo-pastoral area is mainly a place of livestock activities (cattle and small ruminants), the Agro-pastoral area has a mixed farming system where livestock and crop growing are combined. In both areas, agricultural productivity is decreasing, and growing poverty among the poorest is leading to many disparities and inequalities (Maertens *et al.*, 2008; Annabi *et al.*, 2005). As a consequence, rural households are confronted with lower income and mass migration to cities. A description of the households in Chapter 2 shows a diverse ethnicity concerned by the research work and farming systems in Senegal. The SP area is mostly populated by Peulh (73% of the sample) and Wolof (27%) whereas the AP area is inhabited by Wolof and Sereer. Average household size in the AP area is 12 persons compared to 9 in the SP area. The latter area also has a low level of education. In general, women in the two areas show a high level of illiteracy: 81% in the SP, and 68% in the AP area. Moreover, women do not have property rights to land. This issue has a great impact on their income, especially in personal crop-growing.

The agricultural sector consists of cash crops like groundnut, water melon and cotton, to be found mostly in the AP area, and food crops like millet and beans in the SP area. In the latter, farmers are mostly involved in livestock activities like sedentary herders or pastoralists who seasonally move their livestock to a different region. Various forms of extension services in the agricultural sector seek to intensify production. The results in Chapter 3 show that agricultural activities involve all household members. Moreover, women and girls are responsible for daily chores and care of household members. There is a specialization in daily agricultural activities, and women are more productive in cash crops than in livestock, especially in the AP area. In the SP area, women are mostly responsible for milking and selling the milk in the market. Their involvement in farming does not improve their income substantially since the husband's income represents almost 80% of total farming income. The wife's total income averaged 15% of that total. Other sources of income are off-farm activities and cash transfers from migrants. In all cases, men were the main earners in the household.

7.3.2 The allocation of labour and time to farming activities

The allocation of labour within the household is affected by structural and socio-economic factors in the two areas. The description of farming activities in Chapter

3 shows that income procurement varies considerably across zones. While farmers' revenues in the Sylvo-pastoral (SP) area are more likely to depend on livestock activities, households in the Agro-pastoral (AP) area secure their income by mixing crop farming with livestock activities. The diversification of activities observed in the latter area is a consequence of land shortage and land availability. For this reason, livestock and crop farming seem to be complementary and competitive with regard to labour and time allocated to farming. The analysis using Heckman's sample selection model shows that as a result of land shortage in the Agro-pastoral area, keeping small ruminants competes with the growing of crops and has a negative impact on income. Our results show that in crops as well as in livestock, females are more productive than men. This productive participation in agricultural activities is favoured by an effective contribution of girls to housework chores where they act as complementary to their mothers, and in market production where they complement their mothers' activities. Income procurement in the household highly depends on time allocated by females and boys in the care of livestock, and varies across areas. For women, time used in taking care of livestock, savings and cultivated land constitutes an important predictor of their earnings from farming. Time used by husbands in livestock and crops also is a positive predictor of women's participation in both activities. This finding is related to the patriarchal culture in both areas (Sereer in the AP, and Peulh in the SP area) and may imply that husbands impose power over wives regarding their participation in farming activities. In addition, caring for ill members as a household task is negatively associated with farming, especially on livestock, both for husband and wife. The results are economically meaningful and suggest that the time allocated to caring for ill household members decreases the total earnings in livestock farming which is time-consuming.

However, despite their productive contribution of labour to farming activities, results show that the male head of the household is the main controller of land and income from selling livestock and crops. Women mostly receive small revenues from selling cereals or dairy products; these revenues vary across zones. Women in the SP area are relatively wealthy because they own part of the household livestock, because Peulh families give their daughters livestock as a dowry when they marry and move to the husband's place. In all areas, extension programs and assistance from NGO services has induced the emergence of dynamic women's organizations, especially in dairy production (examples are the Women's Directory Board in Livestock in the AP area (DIRFEL) and the Hinger Project in the SP area). These results are consistent with the findings concluding that adult women are more involved in intensified dairy farming than adult men (Tanka *et al.*, 1999; Muriuki *et al.*, 1997; Mullins *et al.*, 1996; Delgado *et al.*, 2001).

Results reveal that education of both husband and wife and household size and ethnicity are significant determinants of crops and livestock activities. All over the sample high levels of men's education positively affect livestock production with a substantial economic meaning. The positive effect of education is especially visible for men in livestock production in the AP area. For women, better education leads to less participation in growing crops, but may increase individual earnings from crops. Household size in terms of labour force participation has a positive effect on farming, and ethnicity matters when it comes to the involvement in either crop or livestock activities. The inclusion of individual health status in the production equation shows that better health induces more participation in crop activities, especially for male heads. For females, no significant effect is observed.

Women's off-farm earnings are positively related to their access to micro-credit. The role of off-farm earnings in mitigating risk from farming is especially observed in the Sylvo-pastoral area where farmers invest in crop production to increase revenues. Wife's savings also positively affect livestock activities in terms of participation and earnings, especially in the SP area. In contrast, cash transfers from migrants induce a negative effect on participation in livestock activities both for men and women, especially in the SP area. This finding confirms that revenues from migration of rural people have a negative impact on farming in both areas and decrease the participation in agricultural activities. This result points at the important role of money transfers in the rural economy (Fall *et al.*, 2004) and confirms the increasing importance of income from non-agricultural activities in rural settings (Ndiaye, 2005). The implication is that an increase of migration away from rural areas would probably decrease agricultural activities. In fact, cash transfers from migrants, instead of being reinvested in agricultural production, are used to resolve other needs of the household (housing, water access). Finally, as expected, it seems that infrastructure like roads is beneficial to marketing activities. The distance from the household to the nearest peri-urban or urban market is a positive predictor of higher male or female earnings.

7.3.3 Decision-making power in rural households

In rural households in developing countries, the male head is the key decision maker when it comes to resource allocation and consumption. In fact, decision-making power in rural households affects the status of household members (especially women) and may be a threat to the well-being of the household members. Differences in cultural behaviour and human capital may affect women's power and their participation in the management of the household. Chapter 4 tests the hypothesis that decision-making towards expenditures is gendered and that the status of women in terms of decision-making regarding food provision, health decisions and children's schooling is central to the well-being of the household. Women's power and important factors affecting female decision-making in the household are shown by descriptive analysis and factor analysis. The results reveal that men more than women are responsible for providing food and managing finance. Regarding decisions about health treatment and health care, men are the exclusive decision makers. On the other hand, females are the main care-givers for ill members of the household.

The comparison between areas shows that SP women have more voice than AP women in decision-making concerning the purchase of food and the management of household finance. Women's weak position in decision-making contrasts with the findings in Chapter 3 that revealed that women allocate as much time as men to farming activities. The latter result also corroborates the role of women in health care given to ill members of the household. This finding confirms the gender division of labour where women are confined to the domestic sphere (Niehof, 2004; Kabeer, 1994). The relative status of women was associated with their mobility outside the household and their voice in food provision and finance management. These factors are highly associated with the age of women, their financial contribution to food provision, their non-labour income and their assets. These results suggest that experience, assets and capital endowments make women more powerful in contributing to household well-being. The access to micro-credit and the cash money transferred to women appear to be important factors of women's power. The positive association has economic and substantial meaning. The fact that capital and asset endowments constitute important determinants of women's voice and empowerment in decision-making regarding expenditures, suggests that policies towards wealth creation should be emphasized as well as policies aimed at increasing especially female livestock assets. Moreover, the study shows that 'asking their husbands' permission for mobility purposes' (going to the market and to the hospital, to buy medicine for children) is negatively associated with women's power. The constraint of 'asking permission' varies across zones and is more visible in the AP area. However, wife's savings and wife's access to micro-credit have a negative impact on these constraints and make women more involved in the labour market. The finding supports the evidence that policies aiming at empowering women through supportive projects have a positive impact on children's health and well-being (Meherun, 2006).

7.3.4 Bargaining relationships between husband and wife and their effects on consumption

Chapter 5 showed the gendered role of husband and wife in consumption, and then analyzed the hypothesis of difference in preferences between husband and wife with respect to the consumption pattern. In the second part, using a non-cooperative household decision model, Chapter 5 tested the hypothesis that husband and wife behave differently towards household and private consumption goods.

Despite their low income described in Chapter 3, women contribute to all household expenditures including those on health and schooling. These contributions are more visible for major foods consumed and clothes (15% and 36%, respectively, and amounted to 13% for schooling and 14% for health expenditures. The wealthier position of SP women may explain their high contribution in consumption goods compared to AP women. These findings suggest that women's participation in the well-being of the household increases with their financial position. The test of income pooling rejects the unitary model of the household (Becker, 1981) assuming the maximization of only one utility function. The results show that husband and wife do not pool their income spent on health or clothes. However, expenditures on food and school are made in a consensual way. The rejection of the unitary model partly confirms empirical findings in developing countries showing that consumption priorities differ across spouses and that female and male income is allocated differently (Quisumbing and Maluccio, 2000; Haddad *et al.,* 1997). One important implication is that policies should target both spouses and be designed according to husband's and wife's needs. Food expenditures vary positively and significantly both with the wife's and the husband's income. While additional income of males is more likely to be spent on household goods (health expenditures for household members) and private goods (clothes and other personal goods), additional income for women increases significantly expenditures in schooling for children.

Therefore, the results lead to the conclusion that husband and wife have different priorities in the consumption of goods (Phipps and Burton, 1998; Maurique and Jensen, 2008). Besides, the contribution of spouses to consumption varies with respect to area and socio-cultural characteristics. For example, it has been observed that in the SP area, expenditures on food vary significantly only with the wife's income. The same is noticed for health and clothing. However, in the AP area, expenditures on clothing for women and children vary only with the wife's income. These contributions are affected by structural and other socio-economic factors. Apart from the household size which is a good predictor of consumption, the other important determinant is again the wife's capital endowment via her savings which provide the ability to smooth consumption. Such findings are observed elsewhere in other studies (Smith, 1999; Kong *et al.,* 2005) and reinforce the arguments that policies should improve and induce women's wealth. For this point of view, the priorities defined in the Poverty Reduction Strategic Paper in Senegal (PRSP) (IMF, 2006) in order to improve women's economic status, are highly relevant and topical. Our results also brought out the important role of the Catholic Church in household investment in health-related expenditures via the indirect effect of extension programs.

The adoption of the Cournot model assuming heterogeneity of preferences of husband and wife tests the assumption that decision-making results in a cooperative outcome where each spouse tries to maximize her or his own well-being taking the partner's behaviour as given. Results show that men tend to attach more value than women do, to private expenditures (clothes and other personal goods) versus household goods. The results suggest that even with their larger share in income, men have a higher share in the consumption of private goods, whereas women primarily attach high importance to household expenditures (on food, schooling, health of all family members). At the equilibrium, with compensation from husband to wife, women far more than men, value their consumption of household goods. However, in the SP area, women as well as men tend to value equally the private goods because of their comfortable economic position. Moreover, an important factor that may affect the equilibrium values of the parameters is related to ethnicity. Therefore, the results provide evidence that preferences are associated with cultural and community's characteristics

The husbands' monetary compensation to their wives is negatively associated with the women's time allocated to livestock earnings which constitute a significant indicator of their earnings. Also, the area matters in explaining the compensation: the compensation increases in the AP area where also, pooling is dominant. Also, their responsibility in managing finance has a positive effect on the monetary compensation. This finding may explain the fact that women's participation in managing finance is a strong indicator of the wife's power (see Chapter 4). The results may suggest that when women are responsible for the management of household finance, they gain more from their husbands probably because they give priority to spending on household goods. Such findings may be related to their role as caregivers and to socially binding norms to which women are confined in many societies, and which may constitute their own appreciation of welfare (Sen, 1990). In fact, the effect that additional transfers of income to women have on household well-being is consistent with other empirical studies (Lundberg *et al.*, 1997; Browning *et al.*, 1994).That means that women's own well-being is strongly linked to the well-being of all household members.

7.3.5 Subjective welfare and well-being

Subjective welfare and its applications

Having described in Chapter 5 that husband and wife have heterogeneous preferences for the consumption of goods and do not behave similarly, Chapter 6 provides evidence that the Leyden approach can be successful in analysing welfare and well-being in developing countries. Subjective welfare through the Leyden approach is based on individuals' evaluation of income, showing that welfare functions differ between individuals. Differences observed between individuals can be correlated with observable individual or environmental characteristics. This part provides answers that aim to find out whether the partners have different opinions with respect to household income and minimum needs. Our results brought out that husbands and wives were able to answer both the income evaluation question (IEQ) (Van Praag *et al.*, 1999) and the minimum income question (MIQ). In addition, the responses to the IEQ and MIQ have been correlated to other variables, such as household size, age, education, time allocation, gender and decision-making of the spouses, and current income. Furthermore, the welfare function, the poverty line and the subjective poverty line have been estimated for the 600 respondents in our sample.

The average household income needed to make ends meet is higher for men than for women. Results show that men need more income than women do, to feel they are at the same level of welfare. For men the average monthly income that may be sufficient for the household amounted to around 151,000 CFA (1 US$ = 475 CFA during the year of the survey), while for women, the average sufficient income is around 95,000 CFA per month. Significant differences are also noticed for these values across zones. The levels of sufficient income and good income reported for husband and wife in the SP area are higher than in the AP area. These differences may be explained by the more comfortable situation of people in the SP area (who are wealthier due to their larger livestock). The estimated average value of the want parameter representing the location of the welfare function is higher for husbands $\mu(hxu)$ than for wives $\mu(wxu)$, 11.62 compared to 11.20. The subjective economic welfare values of their own incomes are 0.72 and 0.73, respectively, for husbands and wives. The results suggest that men as the main providers of consumption goods in the household are less satisfied than women with the same income. Taking into account the six welfare levels used, the individual welfare function shows reliable estimates with average R^2s of 0.93 for husbands and 0.94 for wives. With respect to the minimum income question (MIQ), our results show that husband and wife do not agree on the minimum income needed in the household. This finding might be related to the appreciation of the cost of household goods; females may buy some household goods at a lower price than men, especially some of the cheapest goods such as ingredients (spices, vegetables). However, purchases of rice, water (for ruminants), oil or cereals are more expensive, and are handled by men. In the SP area, expenditures related to water are very costly especially during the dry season. In most cases the related costs are paid by men. The average values reported by husband and wife as a sufficient income to cover household needs, are higher than the average monthly incomes in our surveys. These results mean that, on average, people earn less than the income they reported as sufficient.

Our results showed that for husband and wife, an increase in household or individual income leads to higher needs. While for the husband an important factor in decreasing subjective welfare, was expenditures on private personal goods, for women significant determinants of the individual welfare-function parameter were men's time allocated to crops, household size and decision-making power. Allocation of the husband's labour time to crops tends to make women happier, suggesting that earnings from crops increase women's economic welfare – probably through the financial compensation described in Chapter 5. As we expected, our results suggest that women in larger households are less happy; for the income in such households has to be shared among more people, leaving each worse off than in a smaller household. Regarding decision-making, the findings suggest that women's involvement in it is an important component of women's welfare in Senegal. For the minimum-income question (MIQ), an important determinant was the area; women in the AP area report higher

MIQ values than those in the SP area, suggesting that within the same gender group minimum needs vary across communities and physical environments.

The estimation of the Leyden poverty line (LPL) and the subjective poverty line (SPL) provided estimates of the values describing poverty according to the Leyden approach. While the SPL presented almost the same average value for husband and wife, estimates of poverty according to the LPL showed that for males, poverty lines are much higher than for females. These findings confirm the fact that men's needs are higher than women's. These estimates also vary across the AP and SP areas. For the LPL and the SPL, the AP area showed higher values than the SP area, suggesting that AP farmers feel poorer than SP farmers. Further, the subjective poverty lines showed higher values for better-educated individuals, for households whose women have access to credit and whose farmers have access to off-farm jobs. However, a positive significant effect was only found for women's access to credit. This means that opportunities or policies aiming to increase people's earnings tend to increase their needs compared to others with the same level of income. That means that the subjective poverty lines depend on income and earnings, and this also suggests that men feel poorer because of their higher needs compared to women with the same income. Results indicated that people who benefit from credits need more income than their counterparts who do not benefit from these opportunities in order to reach the same welfare level. In terms of policy implications, the LPL and the SPL present relevant results in estimating poverty in developing countries. These methods can be applied in different situations to evaluate the welfare level and to provide suitable policies. However, further improvements may include the identification of points where we can distinguish between the transient poor who can recover from a shock without help, and those who face chronic poverty (Ndirangu, 2007; Barrett, 2005).

Determinants of subjective well-being

Subjective well-being is a wider concept than welfare and describes the quality of life. The estimated equations comprise important aspects of life including health status and work satisfaction. Husband and wife separately answer the question comprising seven qualification levels, ranging from worst to best possible life. Regarding their life, their health and their economic activities, most respondents perceived to be at the mid-level of well-being. Overall, women were less satisfied with their situation than men. For example, for the economic activities, 19.8% of men feel they are in a good situation compared to only 6% of women. The results of the estimation of the well-being equations showed significant differences between the two areas: individuals generally are happier in the SP area. The well-being in that area is associated with income and, as we saw in Chapter 5, farmers are wealthier there.

As expected, subjective well-being is positively predicted by husband's income. The effect was positive and significant for the husband; for women, no income effect was noticed. This finding supports reasoning in the direction that the answers to the well-being questions used in our study allow for interpersonal comparisons, meaning that people living in the same household may appreciate their well-being status differently (Plug, 1997). For wives, the multinomial probit estimation showed that their responsibility in food provision and their mobility predict their subjective well-being. Asking for permission negatively predicts women's well-being. Household size predicted well husband's well-being since it leads to more happiness. This effect may be explained by the side-effect of labour supply that children may affect to farming activities and also, in most African countries, children are part of well-being. Age did not seem to have any effect on subjective well-being in Senegal. However, a study in Tanzania found decreasing subjective well-being of women older than 60.5 years (Dimoso, 2009). Other background variables, such as the area, are important: AP area predicts unhappiness both for husbands and wives. This result confirms our argument described in the previous section, and is in line with the results in Chapter 3 reporting that with respect to income procurement the AP area has less economic power than the SP area.

Lastly, we may conclude that the Leyden approach improves the materialization of the utility concept in the measurement of welfare and well-being. In fact utility has always been a key issue for economists in interpreting happiness. The Leyden method estimates the utility level by putting direct questions to respondents as to how happy they are with their income and with their current situation. Our findings brought out that the Leyden approach can be successfully applied in the context of developing countries. Without any difficulty, respondents in the different communities answered the IEQ and MIQ as well as the Cantril question. The inclusion in the welfare and well-being equations of background variables such as polygamy, religion, ethnicity, and decision-making, has improved the significance and the relevance of the pseudo R^2. In most cases individual income is taken as the most important reference in the estimation of individual welfare; in the estimation of utility the equations estimated in the Leyden approach have made use of economic and social variables. For this reason, results are relevant to a better understanding of the utility concept. However, we may recognize the difficulty in the choice of relevant variables that may correctly reflect the social environment (binding norms, customs, and values) in which people are embedded. The results also provide additional evidence that in families with two earners, partners do not appreciate their income in the same way. These results are in line with the work of Plug (1997) in German families with two earners, and deviate from the Leyden School results that mostly assume that men and women have the same evaluation of their family income (Van Praag, 1968, 1971; Kapteyn *et al.*, 1988; Rainwater, 1974). Finally, the concept of well-being expanded our research beyond the income criteria, and made use of other important notions of our life, such as health and family life. In terms of policy applications, our results show the relevance of using the Leyden approach in the estimation of welfare and poverty levels. In the context of West-African countries, the method may be useful in the assessment of the Strategic Document for Poverty Reduction implemented in Africa in the 1990s. Also, the Leyden approach may be easily used in the assessment of recent projects in Senegal focused on improving farmers' income, such as the *REVA Plan* (Plan to return to agriculture in Senegal) and *GUANA* (Great Agricultural Offensive for Food and Abundance).

7.4 Relevant findings with respect to difference between areas

The results presented in the previous sections show that in one and the same country, diversity of land use, farming systems and socio-cultural factors may lead to variability in income and livelihoods that potentially affect the gender relations in households. We discovered that while farmers in the SP area draw their revenues from livestock products, households in the AP area do so from selling cereals and from trade. The results show that off-farm earnings have a great impact on the mitigation of risks, especially in the SP area, where these revenues make farmers more sedentary through investment in crops. Also, different distances to the nearest urban centres and markets negatively affect activities related to trading agricultural products (especially crops and milk), particularly in the SP area. With respect to education, an important factor in enhancing productivity, the results show contrasting findings: while men's education induced more earnings in livestock activities in the AP area, it is negatively associated with females' earnings from crops in the AP area. However, considering the entire sample, the wife's education has a positive effect on crop earnings. In the SP area, there is no significant effect of education on either crop or livestock earnings.

The main finding is that husbands are the chief breadwinners, despite the high contribution of females in farming in the two areas. In both areas, women receive some monetary compensation for their work from their husbands. In the AP area they work mainly in cash crops; in the SP area mainly in livestock. In the latter area women gain more than in the former. The results show that in the SP area their livestock gains give them more voice in decision-making regarding their mobility, the purchase of food and the management of finance. These findings suggest that women's earnings are an important factor in their power on decision-making. Constraints relative to their mobility (going to the market or to the hospital) negatively impact on their power and are more visible in the AP area. However, capital endowment such as wife's savings and wife's access to micro-credit, negatively affect this constraint. As for decision-making in consumption, household expenditures on health, food and clothing, in the SP area these vary significantly with wife's income, while in the AP area, women's income impacts only on clothing.

In the SP area, assets and earnings are more individualized than in the AP area. For example, the difference between areas is visible when it comes to the way of spending household income; pooling income was found to be dominant in the AP area only. In the SP area individual spending of earnings (no pooling) was predominant. This could suggest that when women's incomes rise (as in the SP area), income is pooled less and expenditures become more individual. The important point is that, in spite of SP farmers being more conservative (in terms of norms and practices) and less educated, women have more voice in decision-making and contribute much more in expenditures than those in the AP area. This is a very interesting finding. It shows how important women's actual earnings are to their power, and that cultural norms about female submissiveness are relative to what is actually happening in practice. There is also a cultural explanation for this (see also Boutrais, 1984). In fact, Peulh women can enhance their role in life because they are given livestock as dowry by their family when they marry and move to the husband's place. These assets allow them to have their own independent means in the household.

Consequently, the appreciation of welfare and well-being shows differences across the two areas. Compared to the AP area, the SP area shows higher values of welfare estimates (IEQ and MIQ). Furthermore, in the analysis of well-being, people in the SP area feel more satisfied with their life than those in the AP area. Because of the differences between the areas in terms of resource allocations, earnings and decision-making, policies that aim at increasing the welfare of farmers should target people with respect to their needs and constraints, and should take into account the differences in gender relations.

7.5 Relevance of the findings and policy question implications

In Senegal, the major focus of the government is to alleviate poverty by improving income procurement. In the past decade, important programs have been implemented in order to increase income procurement and to reduce poverty. These programs, especially the *REVA* and the *GUANA,* are in line with the vision of the President who believes that people and the private sector should return to agriculture in order to make agriculture the pillar of the economy. These programs tend to promote sustainable development, to limit the migration from rural towns to cities, especially of young people and women, and to increase agricultural production through intensification and diversification by means of better seeds and other inputs. Our research provided empirical results (1) in identifying factors that may contribute to food security, to poverty alleviation and increased earnings for smallholder producers in Senegal, (2) in providing better understanding of intrahousehold dynamics and decision- making with respect to food provision and consumption, and important determinants that impact on decision-making, (3) analyzing household coping strategies with respect to health problems and related expenditures, (4) determining factors that may affect welfare and well-being of individuals in Senegalese households. The results of our study have the potential to provide decision makers with strategic information to make informed choices in household developmental policies with respect to agricultural activities, poverty reduction and health. More specifically, in terms of policy implications, the results suggest that understanding decision-making processes concerning resource allocation and expenditures, is essential in avoiding policy failures. For example, the choice of programs used for poverty alleviation should take into account heterogeneous preferences of the spouses, for example, in the domain of health. In contrast, households may be targeted as a whole in the domain of education or food provision.

In the case of public transfers from the government, targeting an individual spouse will also have gender-specific consequences. For example, we may expect that direct transfers to women will have a positive effect on children's schooling, especially in the AP area. Moreover, taking into account agro-ecological differences, one may expect that direct transfers to women or projects reinforcing women's involvement in the labour market, will induce better health, decrease illiteracy and even improve food provision, for example in the SP area. Besides, projects aiming at capital accumulation or reallocation of resources within the household through the wife's savings, may achieve the same goal by reinforcing women's capability with a direct impact on health and the provision of private expenditures, for instance on clothing.

Policies aimed at improving women's empowerment in these areas should target particularly these aspects: livestock production, land ownership, capital endowment (access to micro-credits) and health issues. For example, policies aimed at providing more incentives in the domain of livestock activities will affect women's valuation of private goods and increase their share in expenditure on personal needs. Moreover, the economic empowerment of women through livestock earnings and access to micro-credits would indirectly increase their savings that are an important element of women's earnings (Chapter 3). The positive effect of land ownership on women's share of household expenditures shows that more assets for women imply a beneficial impact on the behaviour of the household. This would lead especially policy makers to pay attention to property rights. Finally, as we can see, the responsiveness to the valuation of consumption goods may significantly be affected by health problems, especially for women. The evaluation of minimum income brought out that farmers did not reach the income they considered sufficient for the household to make ends meet, suggesting that policies should aim at improving farmers' income through subsidies of certain inputs to improve agricultural productivity.

The final beneficiaries of this research will be the farmers. Implementing suitable policies and programs (access to resources, credit, supportive projects and programs aiming at capital accumulation), will probably assist men and women in the management of their activities. The issue of health may have implications for poverty-reduction policies in the agricultural sector. For instance, the identification of vulnerable people and decision makers regarding health problems (care, costs of treatment) in the household could guide appropriate political strategies concerning access to health facilities and targeting vulnerable people. For example, the creation of subsidized health centres much nearer to the villages may improve the access to health care and treatment.

7.6 Limitations of the research and recommendations for further investigation

Agricultural and off-farm activities in Senegal are affected by a number of constraints, important ones being the irregularity of rainfall in the rainy season, and the lack of inputs. Moreover, most farmers are confronted with low production or bad outcomes and prefer to migrate to urban areas looking for jobs. Consequently, production and income may vary over the years. Our data set is based on cross-sectional data for one year. Observations over several consecutive years (panel data) may provide more insights into the factors affecting farmers' earnings and decision-making over time, and labour in case of income variation. But the limitation of our research field to one year did not permit such investigations. Further improvements will need longer periods of observation in order to include other determinants that may affect intrahousehold dynamics, like inflation in consumption goods, variability in rainy seasons and other factors.

The evaluation of income was based on the report of all agricultural and non-agricultural activities and also on transfers of income, but the application of the questionnaires may suffer from a misevaluation of certain income-procurement activities. For example, women may underestimate their income from small trade in marketing agricultural or non-agricultural products, since market earnings are generated daily and our questionnaire was retrospective, in certain cases. The same problem may arise in the case of men occasionally selling livestock (especially small ruminants). In fact, for farmers this occasional activity is a form of strategy for mitigating shocks if there is not enough money to meet daily food or emergent health costs. A daily or weekly survey will be more appropriate in reporting all earnings. However, in the context of rural African societies, it can be difficult to report all earnings since there is no logbook for income and costs related to household activities. Databases employing large numbers of surveyors and having sufficient funds for the daily collection of data may provide additional relevant statistics.

The identification of social variables that reflect the ethical and social norms of communities concerned was relevant. We included polygamy, religion and ethnicity as social and cultural factors. However, the group discussions and the literature review on the social background of the communities concerned have helped in discussing our results.

The Senegalese government has recognized the low productivity of the agricultural sector and the importance of socio-economic factors in improving production and income. For this reason, the ongoing activities within the Poverty Reduction Strategy Document (PRSD) have focused more on the micro level with special attention to gender issues, health problems and education (IMF, 2006). In the specific domain of agriculture, some other important projects and programs aim at intervening at the micro level (REVA Plan and GUANA) to improve income and well-being (intensification, diversification and credits). The following aspects including micro- and macroeconomic issues may be considered for further areas of investigation.

For example, at the micro level, the extension of the research to other vulnerable people like widows or isolated families with only one head, either male or female, may be considered, in order to draw founded conclusions about factors affecting total earnings and decision-making in rural households. Moreover, it is necessary to extend the investigations to urban and peri-urban locations with a focus on data related especially to income-procurement activities. These investigations should include migrants and may improve the insights regarding the impact of income on welfare and poverty level of their rural families. The current policy reforms also comprise the emergence of intensified farms and agricultural enterprises in dairy and crop systems financed by the private sector. Due to variation in farm size, specialization, actors involved and location, it will be relevant to investigate the impact of different factors on productivity, income and welfare. In-depth analysis can address these issues by taking into account farm structure, people involved and specialization of farms.

In the domain of health issues people afflicted by HIV/AIDS were hard to find because of the low prevalence of the disease in Senegal (0.7%). However, when we consider all health problems together, diseases may disadvantage rural people through inequality in access to resources and infrastructure. This could merit an in-depth analysis of factors that may decrease inequality and poverty. Finally, strong improvements should be made in the identification of social indices (attitudes, norms) that can affect decision-making and women's empowerment in rural households. In fact, such understanding may focus on social backgrounds that can explain stigmatisation in health issues.

The assessment of micro-economic factors should be completed by studying the impact of meso- and macro-measures such as services, prices of inputs and direct interventions in extensions and subsidies. For example, analysis should be extended to valuate chain-organization systems in order to identify the institutional determinants of productivity.

Appendices

Appendix 1. Questionnaire to be addressed at the level of the household[9]

Region:
Zone:
Village:
Name of household:

Religion of chief of the household:
/_/ Muslim
/_/ Catholic

Ethnicity of the household:
/_/ Peulh
/_/ Wolof
/_/ Sereer

Date of interview:
Name of the interviewer:
Distance to weekly market:
Distance to central market:

Name of the respondent:
Size household:

Yields	hectares			
Yields cash crops:	water melon	cotton	peanut	
Yields food crops:	maize	millet	bean	sorghum

Number
Number of goats
Number of sheep
Number of horses

[9] Intrahousehold resources allocation and wellbeing in rural Senegalese household: approved PhD at WU/Fatimata Dia Sow.

General information (concerning persons living in the household)

No (1)	Name (2)	Sex (3)	Age (4)	Status (5)	Relation with head (6)	Level of education (7)	Type of work (8)	Other activities(9)	Current health status (10)
1									

(1) head = 1
(2) name of members of the Household
(3) male or female (1 = m; 2 = fe)
(4) age
(5) single = 1; maried = 2; divorced = 3;
(6) wife = 1; parent = 2; child = 3; grandchild = 4; cousin = 5; niece/nephew = 6; aunt/uncle = 7; other kin = 8
(7) level: primary school not finished = 1; primary school finished = 2; secondary school not finished = 3; secondary school finished = 4; not alphabetized = 5; alphabetized = 6; not applicable = 7
(8) not applicable = 1; school/study = 2; employed = 3; self-employed = 4; unemployed = 5
(9) not applicable = 1; farmer = 2; fisherman = 3; trader = 4; entrepreneur = 5; shopkeeper = 6; teacher = 7; employee = 8; other = 9
(10) healthy = 1; ill = 2

Supporting Organization in the village (NGO, governmental services.)
1
2
3
4
5

Form of services allocated by these organizations
1 = formation
2 = extension of technologies
3 = credits
4 = support in management
5 = other

Type of technologies concerned
1 = feeding strategies
2 = crossbreeding
3 = management of credit
4 = marketing
5 = techniques of processing
6 = organization of farmers
7 = other

How far is the livestock extension center in this community?

Did you have to pay for services allocated by extension services?
/_/ yes
/_/ no

Name of the organizations which support the villages in term of credit or micro-credit.
1 =
2 =
3 =
4 =
5 =

Does the credit program supply loan for specific purpose?

Who operate the credit program?
1 = government
2 = private bank
3 = money lender
4 = cooperative
5 = other

1. Gender division of labour in productive activities (questions should be asked separately to husband and wife)

1.1. Labor and time participation in livestock: estimated average (hours/day) per person.

Livestock activities	Husband	Wife1	Wife2	Boys	Girls	Other
Marketing milk						
Guarding cattle						
Guarding small ruminants						
Milking						
Watering						
Supplementing						
Marketing cattle and small ruminants						

1.2. Labor and time participation in agriculture (estimated hours/day) per person.

	Peanut		Cotton		Cereals	
	Wet season	Dry season	Wet season	Dry season	Wet season	Dry season
Husband						
Wife 1						
Wife 2						
Adult Girls						
Adult Boys						
Hired labour						
Other (sourga)						

1.3. Domestic work (hours/day/person).

Activities	Husband	Wife1	Wife2	Boys	Girls	Other
Repair/fixing						
Fetching water						
Cooking						
Care of children						
Cleaning						
Fetching wood						
Care of ill person						
Kitchen garden						
Other						

1.4. Off-farm activities (breadwinner orientations)

1.4.1 Complete the table.

	Husband*	Wife 1*	Wife 2*	Boys*	Girls*	Other*
Employee in city						
Craft industry						
Mason						
Metalworking, …						
Repair work						
Agricultural labour						
Homemade						
Other						
Total 1						

*Precise the number of months in 2006 and the income generated.

1.4.2. Ask the wife:
In case you don't work, explain why?

1.4.3. Ask the husband:
In case you don't work, explain why?

1.4.4. If you would have extra money what do you prioritize to spend it on?
Ask the wife:
1
2
3
4
Ask the husband:
1
2
3
4

1.4.5. Importance of work: (ask separately).
Ask the husband:
Whose job would you say is considered more important in your family?
/_/ husband's work is considered more important;
/_/ wife's work is considered more important;
/_/ both's work is considered equally important.
Why?

Ask the wife:
Whose job would you say is considered more important in your family?
/_/ husband's work is considered more important;
/_/ wife's work is considered more important;
/_/ both's work is considered equally important.
Why?

1.4.6 Ideally, who do you think should be the financial provider for your family?
/_/ husband only
/_/ husband more than wife
/_/ husband and wife equally
/_/ wife more than husband
/_/ wife only
/_/ should not matter who is provider

2. Gender division of source of income (year 2006)

2.1. Cash crop

Total cash crop sold: cotton	kg	Price/kg	CFA		
Total cash crop sold: peanut	kg	Price/kg	CFA		

Quantity of cash crop sold by man: cotton	kg	Price/kg	CFA
Quantity of cash crop sold by man: peanut	kg	Price/kg	CFA
Quantity of cash crop sold by woman: cotton	kg	Price/kg	CFA
Quantity of cash crop sold by woman: peanut	kg	Price/kg	CFA

2.2. Food crop

Total food crop sold (cereals)	kg	Price/kg	CFA
Quantity of food crop sold by man	kg	Price/kg	CFA
Quantity of food crop sold by woman	kg	Price/kg	CFA

2.3. Dairy

Quantity of fresh milk produced per day in the household	liters
Quantity of sour produced in the household per day	liters

Quantity of fresh milk consumed in the household/day	liters
Quantity of sour milk consumed in the household/day	liters

Quantity of fresh milk sold by men	liters	Price of fresh milk	CFA
Revenues of men from fresh milk/day	CFA		
Part of income allocated to women/day	CFA		

Is it an obligation or a consensus? Explain

Quantity of fresh milk sold by women	liters	Price of fresh milk	CFA
Revenues of women from fresh milk	CFA		
Part allocated to men	CFA		

Is it an obligation or a consensus? Explain

Quantity of sour sold by man	liters	Price of sour milk	CFA
Revenues of man from sour	CFA		
Part allocated to women	CFA		

Is it an obligation or a consensus? Explain

Quantity of sour sold by woman	liters	Price of sour milk	CFA
Revenues of women from sour milk	CFA		
Part allocated to men	CFA		

Is it an obligation or a consensus? Explain

2.4. Livestock
Ask the husband:

Number of cows sold in (2006)	Value in CFA	
Number of sheep	Value in CFA	
Number of goats	Value in CFA	
Total income		CFA (2006)
How much money is for your wife in this value?		CFA
Did you share revenues with your wife?		
How much did you allow her?		CFA

Ask the wife

Number of cows sold by women		Revenues	CFA (2006)
Number of sheep/goat/		Revenues	CFA (2006)
Share for you husband	CFA		
Share for you	CFA		

Did you cultivate vegetables in 2006?
How much did you gain in total? CFA

Transfers of money from other places or persons: who did the transfers?
Ask the husband: amount transferred to him: CFA
Ask the wife: amount transferred to her: CFA

Summary table of earnings (2006).

	Husband*	Wife 1*	Wife 2*	Boys*	Girls*	Other*
Selling cattle						
Selling goats/chicken						
Selling sheep						
Milk						
Cash crop Peanut Cotton Beans Bissap Water melon						
Food crops Maize Millet Beans Home vegetables						
Trades						
Home garden vegetables						
Other income (from children or relatives in the city or from immigration)						
Other						
Total 2						
Total 1+2						

* Precise amount of income, where and who earns the money.

Total household income CFA

3. Costs of production: intended for the husband and wife in the household

3.1. Livestock
3.1.1. Feeding strategies
Natural pasture 1 = yes 2 = no
Forage: quantity/day kg price/kg CFA
1 = cost paid by women CFA
2 = paid by men CFA
3 = not paid but is resource pooling in household
Tops of peanut: quantity/day kg price/kg CFA
1 = cost paid by women CFA
2 = paid by men CFA
3 = not paid but is resource pooling in household
Tops of cereals: quantity/day kg price/kg CFA
1 = cost paid by women CFA
2 = paid by men CFA
3 = not paid but is resource pooling in household

3.1. Concentrated feeds.

Feeds	Quantity used/month* (precise unity)	Price per unity	Feed is paid by women 1 = yes 2 = no	Feed is paid by men 1 = yes 2 = no	Feed is a resource produced in the the household
Scraped groundnut					
Melasse					
Cotton granulates					
Its of millet					
Other					

* The concentrated feeds are used in dry season: precise quantity per month or per week or for all the season.

3.1.2. Costs related to the health of animals:
Give total cost per season if it is known CFA
Otherwise give a range
0-50,000 CFA
50,000-100,000 CFA
>100,000 CFA
total for dry season CFA
total for wet season CFA
1 = cost are paid by women CFA
2 = paid by men CFA

3.1.3. Water
Water is paid?
1 = yes
2 = no
If yes, precise total cost/day CFA
1 = cost paid by women CFA
2 = cost paid by men CFA
3 = cost paid by other CFA

3.1.4. Guarding ruminants
Guarding of cows costs money, how much total cost/day
1 = cost paid by women CFA
2 = cost paid by men CFA
3 = cost shared CFA
Guarding of cows is not paid: who takes it in charge? Precise (children or other)

3.1.5. Wages for dairy production and marketing
Wage paid to family labor/day CFA
Wage paid to hired labor/day CFA

3.2. Growing crops
3.2.1. Inputs in crops.

Input	Quantity paid	Value in CFA	How much of the cost is paid by women?	How much of the cost is paid by men?
Groundnut				
Bean				
Cotton				
Fertilizers				
Pesticides				

3.2.2. Transportation costs for selling the milk/day
What is the distance of the household to the market?
Transportation costs paid CFA
Who pays?
Man CFA
Woman CFA

3.2.3. Tax paid for livestock/year/cow CFA
Who pays?
Man CFA
Woman CFA

3.2.4. Existence of credit/year
/_/ yes
/_/ no
Amount of credit CFA
Credit allocated to man CFA
Credit allocated to women CFA

4. Gender control over household income and allocation

4.1. Do you share your income with your partner?
Ask the husband
/_/ yes all income
/_/ yes most of the income (50-90%)
/_/ yes, small part of the income (10-50%)
/_/ no
Indicate the value that you share with her: CFA

4.2. Ask the wife
/_/ yes all income
/_/ yes most of the income (50-90%)
/_/ yes, small part of the income (10-50%)
/_/ no
Indicate the value that you share with your husband: CFA

4.3. Do children share their income with their parents?
/_/ yes all income
/_/ yes most of the income (50-90%)
/_/ yes, small part of the income (10-50%)
/_/ no

4.4.Who manages finances in the household?
/_/ husband only
/_/ husband more than wife
/_/ both equally
/_/ wife more than husband
/_/ wife only

4.5. Do you think that the management of the finances in your household is good?
/_/ yes
/_/ no, if not, why?

5. Gender responsibilities in food provision for the household

5.1. Do you think that food purchased and produced per day is
/_/frequently not sufficient
/_/sometimes not sufficient
/_/usually sufficient
/_/sometimes more than sufficient
/_/frequently more than sufficient

5.2. What do you do when you do not have enough food?
/_/ we never have problems in food security
/_/ we skip meals
/_/ we use other substitutes of food (explain)
/_/ we sell cattle, small ruminants or chickens if we have them
/_/ we try to get food gifts
/_/ we collect wild foods
/_/ we sell household goods
/_/ we borrow food
/_/ we borrow money to buy food

5.3.Who usually buys food for the household?
/_/ husband only
/_/ husband more than wife
/_/ both equally
/_/ wife more than husband
/_/ wife only

5.4. Who usually borrows food for the household?
/_/ husband only
/_/ husband more than wife
/_/ both equally
/_/ wife more than husband
/_/ wife only
/_/ we never borrow food

5.5 Who usually exchanges food with other households?
/_/ husband only
/_/ husband more than wife
/_/ both equally
/_/ wife more than husband
/_/ wife only
/_/ we never exchange food

5.6 Who usually borrows money to buy food if needed?
/_/ husband only
/_/ husband more than wife
/_/ both equally
/_/ wife more than husband
/_/ wife only
/_/ we never borrow money to buy food

5.7. Please, list the consumption expenditures in the household.

Goods	Husband's contribution*	Wife's contribution*	Joint contribution
1st			
2nd			
3rd			
4th			
5th			
6			
7			
8			
9			
10			
Other food			

* In terms of expenditures per month.

5.8. Other goods.

Other goods	Total expenditures paid	Husband contribution*	Wife's contribution*	Joint contribution
Clothes				
School of children				
Health				
Other				
Other				

* In terms of expenditures.

6. Women empowerment and bargaining power

To be addressed to the wife:
6.1.Number of owned cattle

6.2. Number of owned sheep and goats

6.3. Access to land
/_/ yes
/_/ no

6.4. Existence of other assets in the household
/_/ yes
/_/ no

6.5 Assets brought into marriage
/_/ yes
/_/ no

6.6. Do you generally ask permission to go to the market?
/_/ yes
/_/ no

6.7. Can you bring your children to the hospital without permission?
/_/ yes
/_/ no

6.8. Can you pay for medical cost without permission?
/_/yes
/_/ no

6.9. Can you decide to spend money without permission if you or your children are ill?
/_/ yes
/_/ no

6.10. Do you have access to credit?
/_/ yes
/_/ no

6.11. Do you have some savings
/_/ yes
/_/ no
If so, please mention amount CFA
How did you get the savings? Explain

7. Analysis of welfare and well-being

7.1 Questionnaire to be applied to the head of household:
Total income reviewed here ....

1. What is the minimum income your household needs in this situation to make ends meet?
2. Taking into account your situation, size of your household, farm activities and other activities, what income per month (in CFA) you think is:

Evaluation/income

Very bad	
Bad	
Insufficient	
Sufficient	
Good	
Very good	

Questionnaire to be applied to the female partner:
Review income here ....

1. What is the minimum income your household needs in this situation to make ends meet?
2. Taking into account your situation, size of your household, farm activities and other activities, what income per month (in CFA) you think is:

Evaluation/income

Very bad	
Bad	
Insufficient	
Sufficient	
Good	
Very good	

7.2 Well being

Questionnaire to be addressed to the female head of household:

7.2.1. Here is a picture of a ladder. If the top of the ladder represents the best possible life for you and the bottom represents the worst possible live for you, where on the ladder do you feel you personally stand at the present time?

7. Best possible life
6
5
4
3
2
1.Worst possible life

7.2.2. Consider now your health

7. Best possible health
6
5
4
3
2
1.Worst possible health

7.2.3. Consider your work

7. Best possible work
6
5
4
3
2
1.Worst possible work

Questionnaire to be addressed to the partner:

7.2.4. Here is a picture of a ladder. If the top of the ladder represents the best possible life for you and the bottom represents the worst possible live for you, where on the ladder do you feel you personally stand at the present time?

7. Best possible life
6
5
4
3
2
1.Worst possible life

7.2.5. Consider now your health

7. Best possible health
6
5
4
3
2
1.Worst possible health

7.2.6. Consider your work

7. Best possible work
6
5
4
3
2
1.Worst possible work

Appendix 2. Health and illness problems

I. Questionnaire Euroqol EQ-5D

By placing a tick in one box in each group below, please indicate which statements best describe your own health state today.
Ask the husband:

Mobility
- I have no problems in walking about ☐
- I have some problems in walking about ☐
- I am confined to bed ☐

Self-care
- I have no problems with self-care ☐
- I have some problems washing or dressing myself ☐
- I am unable to wash or dress myself ☐

Usual activities (*e.g. work, study, housework, family or leisure activities*)
- I have no problems with performing my usual activities ☐
- I have some problems with performing my usual activities ☐
- I am unable to perform my usual activities ☐

Pain/discomfort
- I have no pain or discomfort ☐
- I have moderate pain or discomfort ☐
- I have extreme pain or discomfort ☐

Anxiety/depression
- I am not anxious or depressed ☐
- I am moderately anxious or depressed ☐
- I am extremely anxious or depressed ☐

Ask the wife:

Mobility
- I have no problems in walking about ☐
- I have some problems in walking about ☐
- I am confined to bed ☐

Self-care
- I have no problems with self-care ☐
- I have some problems washing or dressing myself ☐
- I am unable to wash or dress myself ☐

Usual activities (*e.g. work, study, housework, family or leisure activities*)
- I have no problems with performing my usual activities ☐
- I have some problems with performing my usual activities ☐
- I am unable to perform my usual activities ☐

Pain/discomfort
- I have no pain or discomfort ☐

I have moderate pain or discomfort ☐
I have extreme pain or discomfort ☐
Anxiety/depression
I am not anxious or depressed ☐
I am moderately anxious or depressed ☐
I am extremely anxious or depressed ☐

2. Care and treatment

2.1. Distance of the local dispensary from the household
Name of local dispensary:
Distance from the household km

2.2. How far is the nearest pharmacy from the household? km

2.3. When the wife is ill, who replaces her in cooking and takes care of children?
/_/ husband
/_/ other wife
/_/ relative
/_/ daughters
/_/ neighbor

2.4 Who decides on the expenditures concerning illness problems?
/_/ husband
/_/ wife
/_/ both
/_/ adult child

2.5. What medication do you usually use in case of chronic illness?
/_/ self-medication (drugs obtained from drugs vendors, street marketers)
/_/ products from healer: traditional medicine
/_/ pharmacy
If you use medicament from a healer, please explain why

2.6. How much time does it take per day to care for the ill person? hours/day.

2.7. What are the reasons of not going to the hospital in case of illness?
/_/ distance
/_/ costs of consultation
/_/ treatment is not effective
/_/ costs of medication
/_/ quality of services are not good
/_/ lack of care and personal
/_/ I don't have time

2.8. If the quality of health services is not good, explain in which way

2.9. Do you have local programs (NGOs or other) that support you in prevention of illness?
/_/ yes
/_/ no
If yes, give precise name of the organization

2.10. Identification of important health problems the household faces
List name and information for the year 2006.

Name of the person	List of frequent health problems during the last 6 months	Please indicate where the treatment was applied (1)	Days of illness	Main care given by(2)	Level of expenditures treatment/drugs	Origin of the money (3)	Opinion about the treatment (4)	Recovery (5)

(1): 1 = at home, 2 = at the hospital, 3 = in a health center, 4 = to the healer, 5 = both hospital and healer, 6 = other)
(2): 1 = daughter, 2 = son, 3 = husband, 4 = relative, 5 = neighbor, 6 = wife1, 7 = wife2
(3): (1 = from the husband, 2 = from the wife, 3 = both, 4 = from abroad, 5 = from relative, 6 = from other
(4): 1 = good, 2 = middle, 3 = bad
(5): 1 = fully recovered, 2 = still ill but almost recovered, 3 = still ill (chronic illness)

2.11. Persons who died during the past year in the household?

Name of the person	Cause of death	Age of death

3. Family planning and HIV/AIDS

Ask the wife:
3.1. What do you think about family planning?
What does your husband think about family planning?

3.2. Do you use family planning?
/_/ yes, if so explain
/_/ no, if not, why?

3.3. Do husband and wife discuss family planning?
/_/ yes
/_/ no

3.4 If not, explain

3.5. Do husband and wife discuss HIV/AIDS?
/_/ yes
/_/ no

3.6.Which knowledge do you have concerning HIV/AIDS (cause, transmission, treatment....)?

3.7.How did you obtain this knowledge?
/_/ through the media (TV, radio)
/_/ through other people
/_/ sensitizing by supportive organizations

3.8. Do you think that in your community, people suffer from HIV/AIDS?
/_/ yes
/_/ no

Appendix 3. Determinants of husband and wife's expenditures (comparison between areas)

Variable	Sylvo-pastoral area							
	Food		Clothes		Health		Schooling	
	Coeff.	t-value	Coeff.	t-value	Coeff.	t-value	Coeff.	t-value
Monthly husband income	0.0373	1.36	0.0560	1.37	0.1081	0.83	-0.003	-0.02
Monthly wife's income	0.0698	4.54***	0.0195	0.85	0.1920	2.63**	0.2754	3.37**
Age husband	0.0062	0.94	0.0186	1.87	0.0232	0.73	0.0357	1.00
Age wife	-0.0115	-1.71	-0.009	-0.89	-0.039	-1.23	0.0338	0.94
Religion [3]	0.1767	0.38	0.0755	0.11	-0.616	-0.28	5.544	2.26*
Polygyny [4]	0.07501	0.60	0.0090	0.05	0.0168	0.03	-0.311	-0.47
Household size	0.0191	1.32	0.0578	2.67**	0.1060	1.54	0.1844	2.39**
Wife's savings	3.90×10^{-8}	0.24	5.59×10^{-7}	2.29*	6.86×10^{-7}	0.88	3.27×10^{-7}	0.38
Wife's access to credit [2]	0.2301	1.71	-0.066	-0.33	1.027	1.61	0.2759	0.39
Wolof [5]	-0.1976	-1.59	0.2108	1.14	1.016	1.73	1.421	2.16*
Peulh [6]			-		-		-	
Education (H) [1]	-0.2617	-2.01*	0.0388	0.20	-0.342	-0.56	1.188	1.72
Education (W) [1]	-0.0828	-0.53	0.0925	0.40	-0.194	-0.27	0.8603	1.04
Constant	9.246	9.91	6.403	4.61	3.164	0.72	-14.0	-2.83
Adjusted R^2		0.23		0.12		0.12		0.25

$P<0.05$, $^{**}P<0.01$, $^{***}P<0.001$.

H = husband, W = wife.

[1] Dummy (0 = lowesr, 1 = highest).

[2] Dummy (1 = yes).

[3] Dummy(0 = Catholic, 1 = Muslim).

[4] Dummy (0 = monogamous household, 1 = polygamous household).

[5] Dummy Wolof (Wolof = 1, Sereer = 0).

[6] Dummy Peulh (Peulh = 1, Sereer = 0).

Agro-pastoral area

Food		Clothes		Health		Schooling	
Coeff.	t-value	Coeff.	t-value	Coeff.	t-value	Coeff.	t-value
0.0448	1.22	0.1504	3.98***	0.155	1.13	0.164	1.30
0.0053	0.43	0.0051	0.35	-0.043	-0.83	0.063	1.30
-0.008	-0.92	0.0075	0.85	0.0030	0.09	-0.0463	-1.43
0.0098	1.03	-0.005	-0.54	-0.005	-0.15	0.0593	1.78
0.4299	1.41	0.0378	0.12	-1.08	-0.94	-1.68	-1.58
-0.0096	-0.30	0.0394	1.22	0.0349	0.30	0.1281	1.18
0.0247	2.01*	0.0300	2.34*	0.0624	1.34	0.1247	2.91**
9.14×10^{-7}	1.44	1.21×10^{-6}	1.82	3.86×10^{-6}	1.61	4.61×10^{-6}	2.08*
-0.2495	-2.0*	0.2157	1.67	0.2248	0.48	0.3588	0.83
0.23642	1.70	-0.338	-2.35*	-0.352	-0.67	0.0082	0.02
0.26241	1.51	-0.373	-2.06*	-0.470	-0.71	0.7379	1.21
0.09614	0.83	0.1530	1.27	0.1420	0.32	0.7070	1.75
0.02164	0.16	0.2820	2.03*	-0.342	-0.68	0.3039	0.65
9.1588	11.33	5.775	6.86	6.820	2.23	5.185	1.84
	0.09		0.19		0.005		0.18

References

Abadian, S. (1996). Women's autonomy and its impact on fertility. World Development, 24 (12): 1793-1809.

Adjamagbo, A., V. Delaunaye, P. Levi and O. Ndiaye (2006). Comment les ménages d'une zone rurale du Sénégal gèrent-ils leurs ressources? Etudes Rurales, 177: 71-79.

Agarwal, B. (1997). Bargaining and gender relations: Within and beyond the household. Feminist Economics, 3 (1): 1-51.

Allais, M. and O. Hagen (1979). Expected utility hypothesis and the Allais paradox. Dordrecht: Reidel.

Alderman, H. and S. Chishti (1991). Simultaneous determination of household and market oriented activities of women in rural Pakistan. Research in Population Economics, 7: 245-265.

Alston, M. and J. Wilkinson (1998). Australian farm women-Shut out or fenced in? The lack of women in agricultural leadership. European Society for Rural Sociology. Sociologia Ruralis, 38 (3): 391-408.

Amin, S. (1997). The poverty-Purdha trap in rural Bangladesh: Implications for women's role in the family. Development and Change, 28 (2): 213-233.

Annabi, N., F. Cisse, J. Cockburn and B. Decaluwe (2005). Trade liberalization, growth and poverty in Senegal: A dynamic micro simulation GGE Model Analysis. CIRPEE Working paper no. 05-12.

Antoine, P. and P. Bocquier (1995). Les familles Dakaroises face à la crise. IFAN-ORSTOM-CEPED, Dakar, 209 p.

Antonides, G., A. Kapteyn and T.J. Wansbeek (1980). Reliability and validity assessment of ten methods for the measurement of individual welfare functions of income. Working paper, Modelling Research Group, University of Southern California. Los Angeles.

Antwi-Nsiah, C. (1993). Household decision-making in Ghana. Presented at the Sixth International Association for Women in Development Forum, Washington DC, USA.

Ao Onoyo, H.C. (2008). Socio economic determinants of productivity and incomes of rice farmers in Udenu State, Nigeria. Journal of Agriculture and Social Research, 8 (2): 37-52.

Apps, P.F. and R. Rees (1997). Collective labor supply and household production. Journal of Political Economy, 105(1): 178-90.

Backer, M.J. and P.J. Joyce (2005). Marriage, Specialization and the Gender Division of Labor, 2005. From http://jjacobsen.web.wesleyan.edu/wps/CSD1101.pdf.

Barrett, C.B. (2005). Rural poverty dynamics: Development policy implications. Agricultural Economics, 32 (1): 45-60.

Basu, A. (ed.) (1992). Culture, the status of women and demographic behavior. Clarendon Press, Oxford.

Batliwala, S. (1994). The meaning of women's empowerment: new concepts from action. In: Sen, G., A. Germain and L.C. Chen (eds.), Population policies reconsidered. Health, employment and rights, pp 127-138. Cambridge, MA: Havard University Press.

Becker, G.S. (1965). A theory of the allocation of time. Economic Journal, 75: 493-517.

Becker, G.S. (1974). A theory of marriage. In: Schultz, T.W. (ed.), Economics of the family. Chicago: University of Chicago Press.

Becker, G.S. (ed.) (1981). A treatise on the family. Cambridge, MA: Harvard University Press.

Benoit-Cattin, M., and J. Faye (eds.). (1982). L'exploitation agricole familiale en Afrique Soudano-Sahélienne. Presses Universitaires de France, Techniques Vivantes, Paris.

Bentham, J. (ed.) (1970(1789)). Introduction to the principles of morals and legislation. London, Athlone Press.

Bitran. R. (1995). Efficiency and quality in the public and private sectors in Senegal. Health Policy and Planning, 10 (3): 271-283.

Blacher, J.G.C. (1978). A critique of international definitions of economic activity. Population Bulletin of ECWA, 14: 47-54.

Bonke, J., and M. Browning (2009). Allocation of expenditures within the household. A new survey. Fiscal Studies, 30 (3), 461-481.

Booth, A.D., L.W. Johnson and J. Edwards (1984). Women, outside employment and marital instability. American Journal of Sociology, 90 (3): 567-83.

Boserup, E. (ed.) (1970). Women's role in economic development. London: Earthscan Press.

Boutrais, J. (ed.) (1984). Le nord du Cameroun: Des hommes, une région. ORSTOM. Office de la recherche scientifique et Technique Outre- mer.

Brown, L., and L. Haddad (1995). Time allocation patterns and time burdens: A gendered analysis of seven countries. IFPRI, Washington DC. Mimeo.

Brown, L.R. and L. Haddad (1996). Time allocation patterns and time burdens: A gendered analysis of seven countries. Mimeo. International Food Policy Research Institute (IFPRI), Washington DC.

Browner, C.H. (1989). Women, household and health in Latino America. Department of Psychiatry and bio behavioural sciences. University of California, Los Angeles.

Browning, M. (2000). The saving behaviour of a two-person household. Scandinavian Journal of Economics, 102 (2): 235-51.

Browning, M. and P.A. Chiappori (1998). Efficient intrahousehold allocation: A general characterization and empirical tests. Econometrica, 66 (6): 1241-1278.

Browning, M., F. Bourguignon, P.A. Chiappori and V. Lechene (1994). Incomes and outcomes: A structural model of intrahousehold allocation. Journal of Political Economy, 102 (6): 1067-1096.

Bushal, S. (2006). Women access to food and land rights. ANWA. From www.asianruralwomen.net.

Caldwell, J.C. (1986). Routes to low mortality in poor countries. Population and Development Review, 12 (2): 171-220.

Caldwell, J.C., and P. Caldwell (1993). Roles of women, families and communities in preventing illness and providing health services in developing countries. In: J.N. Gribble and S.H. Preston (eds.), The epidemiologic transition: Policy and planning implications for developing countries. Washington D.C: National Academy Press, pp. 252-271.

Cantril, H. (1965). The pattern of human concerns. New Brunwick, N.J: Rutgers University Press.

Carter, M. and E. Katz (1997). Separate spheres and the conjugal contract: Understanding the impact of gender-biased development. In: L. Haddad, J. Hoddinott and H. Alderman (eds.), Intrahousehold resource allocation in developing countries: Methods, models, and policy. Baltimore: Johns Hopkins University Press for the International Food Policy Research Institute, 95-111.

Caverieniere, M. and M. Debene (eds.) (1988). Le droit foncier sénégalais. Berger-Levrault.

Chafetz, J.S. (ed.) (2006). Handbook of the sociology of gender. Handbook of Sociology and social Research. Springer Science + Business Media, LLC.

Chen, Z. and F. Woolley (1999). A Cournot-Nash model of family decision-making. Economic Journal, 111: 722-748.

Chiappori, P.A. (1988). Nash bargained household decisions: A comment. International Economic Review, 29 (4): 791-796.

Chiappori. P.A. (1992). Collective labor supply and welfare. Journal of Political Economy, 100 (3): 437-467.

Christensen, L.R., D.W. Jorgenson and L.J. Lau (1975). Transcendental logarithmic utility function. American Economic Review, 65(3): 367-383.

Ciabattari, T. (2001). Changes in men's conservative gender ideologies: cohort and period influences. Gender and Society, 15 (4): 574-591.

Ciccone, A. (ed.) (1994). Human capital and technical progress: Stagnation, transition, and growth. Department of Economics, Stanford University, Stanford, Calif, USA.

Clark, A.E. and A.J. Oswald (1994). Unhappiness and employment. The Economic Journal, 104: 648-659.

Clark, A.E. and A.J. Oswald (1996). Satisfaction and comparison income, Journal of Public Economics, 61: 359-381.

Collins, D., J. Morduch, S. Rutherford and O. Ruthvan (eds.) (2009). Portofolio of the poor: How the world's poor live on $2 a day. Princeton and Oxford: Oxford University Press.

Colosanto, D., A. Kapteyn and J. Van der Gaag (1983). Two subjective definitions of poverty: Results from the Wisconsin basic needs study. Journal of Human Resources, 28: 127-38.

Coppo, P., L. Pisani and A. Keita (1992). Perceived morbidity and health behaviour in a Dogon Community. Social Science and Medicine, 34 (11): 1227-1235.

Cornelisse-Vermaat. J.R., G. Antonides, J.A.C. Van Ophem and H. Maassen van Den Brink (2006). Body mass index, perceived health and happiness: Their determinants and structural relationships. Social Indicators Research, 79: 143-158.

Cross, C. (1999). Social Capital in KwaZulu Nata: A Qualitative Study. Mimeo, IFPRI, Washington, DC.

Danziger, S., J. Van der Gaag, M.K. Tausig and E. Smolensky (1984). The direct measurement of welfare levels: How much does it cost to make ends meet? The Review of Economics and Statistics, 66: 500-505.

Darity, W. (1995). The formal structure of a gender-segregated low-income economy. World Development, 23 (11): 1963-1968.

Datt, G. and D. Joliffe (1998). The determinants of poverty in Egypt. Mimeo, IFPRI, Washington, DC.

David, M. (1991). His, hers, or theirs: An analysis of conjugal economic relations among the Kpelle of Liberia. PhD. Dissertation, Department of Rural Sociology, University of Wisconsin-Madison.

Davis, B. (ed.) (2003). Food, agriculture and rural development, Latino America and the Caraibes. Current and Emerging Issues for Economic Analysis and Policy Research- II. Rome: Food and Agriculture Organization of the Unites nations (FAO), Economic and Social Department.

De Brauw, A. and S. Rozelle (2008). Reconciling the returns to education in off-farm wages employment in rural China. Review of Development Economics, 12 (1):57-71.

Delgado, C., M. Rosegrant, H. Steinfeld, S. Ethui and C. Courbois (1999). Livestock to 2020. The next food revolution. Food, Agriculture and the Environment. Discussion Paper 28. IFPRI (International Food policy Research Institute), Washington, DC, USA; FAO (Food Agriculture Organization of the Unites Nations), Rome, Italy; and ILRI International research Livestock), Nairobi, Kenya. 72 pp.

Delgado, C., M. Rosegrant and S. Meijer (2001). Livestock to 2020: The revolution continues. Paper presented at the annual meeting of the International Agricultural Trade Research Consortium, Auckland, New Zealand, 18-19 January.

Dey Abbas, J. (1997). Gender asymmetries in intrahousehold resource allocation in Sub-Saharan Africa: Some policy implication for land and labor productivity, Chapter 15. In: Haddad, L., Hoddinott, J. and Alderman, H. (eds.), Intrahousehold resource allocation in developing countries: Methods, models, and policy. Baltimore, Md., USA: Johns Hopkins University Press for the International Food Policy Research Institute, 249-262.

Diener, E. and E. Sandvik, D. Gallagher and W. Pavot (1991). the effects of response artifacts on measures of subjective well-being. Social Indicator Research, 24: 35-56.

Diener, E. and E. Sandvik, L. Seidlitz and M. Diener (1993). The relationship between income and subjective well-being: Relative or absolute? Social Indicator Research, 28: 195-223.

Diener, E. and E.M. Suh (1997). Measuring quality of life: economic, social and subjective indicators. Social Indicators Research, 40: 189-216.

Diener, E., E.M. Suh, R.E. Lucas and H.L. Smith (1999). Subjective well-being: Three decades of progress. Psychological Bulletin, 125: 216-302.

Diewert, E. and D. Lawrence (1999). Measuring New Zealands' productivity, Treasury orking paper 99/5, http://www. treasury.govnz/working papers/.

Dimoso, R.L. (2009). Environmental degradation and intra-household welfare: the case of the Tanzanian rural south Pare Highlands. Ph.D Dissertation, Wageningen University. 181p.

Diop, B.M., M. Sene and P. Kelly (1996). Survey of the quality of integrated management of childhood diseases services in health establishment in the Dakar, Kaffrine and Kebemer health districts. Dakar, Senegal Health Assessment DMCI.

Diouf, M. (ed.) (2002). L'endettement puis l'ajustement: l'Afrique des institutions Brettons Woods. Paris, l'Harmattan.

Dollar, D. and R. Gatti (1999). Gender inequality, income and growth: Are good times good for women? Policy research report on gender and development. Working paper series n01. World Bank: Washington DC.

Doss, C. (1996). Testing among models of intrahousehold resource allocation. World Development, 24 (10): 1597-1609.

Doss, C. (1997). The effects of women's bargaining power on household health and education outcomes: Evidence from Ghana. Paper presented at the Population Association of America Annual Meetings, March 27-29, 1997, Washington DC.

Doss, C., C. Grown and D. Deere Carmen (2008). Gender and asset ownership: A guide to collecting individual-level data. Policy Research Working Paper No. 4704. Washington, DC: World Bank.

Doss, C., C. Grown and C.D. Deere (2009). Gender and rural employment: Collecting individual level assets data for gender analysis of poverty and rural employment. Paper presented at the FAO-IFAD-ILO workshop on gaps, trends and current research in gender dimensions of agricultural and rural employment: differentiated pathways out of poverty. Rome, 31th March-2th April.

Dowd, R. and M. Hoffman (2009). Religion, gender and political participation in Sub-Saharan Africa: lesson from Nigeria, Senegal and Uganda. Paper presented at the annual meeting of the Midwest Political Science Association. 67th Annual National Conference, The Palmer House Hilton, Chicago, IL. From: http://www.allacademic.com/meta/p361010_index.html 2009-03-20.

Duncan, T. (1990). Intrahousehold resource allocation: An inferential approach. Journal of Human Resources, 25 (4): 635-664.

Duncan, T. (1996). Education across generation in South Africa. American Economic Review, 86 (2): 330-334.

Duncan, T. and C.L. Chen (1994). Income shares and shares of income. Labor and Population Working paper 94-08. Santa Monica, California: Rand Corporation.

Duncan, T., C. Dante and E. Frankenberg (1997). Child health and distribution of household resources at marriage. Mimeo, Rand, UCLA.

Duruflé, G. (ed.) (1994). Le Sénégal peut-il sortir de la crise? Douze ans d'ajustement structurel au Sénégal. Paris, Karthala.

Elo, I.T. and S.M. Preston (1996). Educational differential in mortality. Unites States, 1979-85. Social Sciences & Medicine, 42 (1): 47-57.

Ellison, C.G. (1991). Religious involvement and subjective well-being. Journal of Health and Social Behaviour, 32: 80-99.

Elson, D. (1991). Male bias in macro-economics: The case of structural adjustment. In: Diane E.(Ed.), Male bias in the development process. Manchester: Manchester University Press, pp.164-190.

Elson, D. (1992). Male Bias in Structural Adjustment. In: Afshar, H. and Dennis,C. (eds.), Women and Adjustment Policies in the Third World. London: Macmillan Academic and Professional, Ltd.

Elson, D. (1993). Gender-aware analysis and development economics. Journal of International Development, 5 (2): 237-247.

England, P. and G. Farkas (1986). Household, employment and gender: A social economic and demographic view. New York. Aldine.

England, P. and B. Kilbourne (1990). Feminist critiques of the separative model of the self: Implications for rational choice theory. Rationality and Society, 2 (2): 156-72.

Fafchamps, M., B. Kebede and A. Quisumbing (2005). Intrahousehold welfare. Poverty research group, working paper GPRG-WPS-044. Oxford University Press. Oxford.

Fafchamps, M. and A.R. Quisumbing (1998). Human capital, productivity, and labor allocation in rural Pakistan. International Food Policy Research Institute. FCND Discussions Papers no. 48. Washington DC.

Fall, A.S. (1997). La pauvreté rurale au Sénégal: Diagnostic et stratégies de lutte. Banque Mondiale, Groupe de Réflexion Stratégique (GRS), rapport final. Dakar, Sénégal, 161 p.

Fall, A.S., O.S. Sy and M. Sall (2004). Household economies in West Africa, Les economies domestiques en Afriques de l'Ouest: Synthèse régionale. In: A.S. Fall and O.S. Sy (eds.), Household economies in West Africa, Les économies domestiques en Afriques de l'Ouest.: CODESRIA/OXFAM GB, pp. 264-327.

Fall, A.S., A. Gaye and A. Ndao (2004). Les Economies Domestiques au Sénégal. In: Fall, A.S et Sy, O.S. (eds.), Household economies in West Africa. Les économies domestiques en Afriques de l'Ouest. CODESRIA/OXFAM GB, 264-327.

FAO (2003). Measuring impacts of HIV/AIDS on rural livelihoods and food security. FAO Fact Sheet. Rome, Food and Agriculture Organization of the United Nations, Rome.

FAO (2002). HIV/AIDS, Food security and rural livelihoods. FAO Fact Sheet. Rome, Food and Agriculture Organization of the Unites Nations, Rome.

FAO (1998). Situation of food and agriculture. Food and Agriculture Organization of the Unites Nations, FAO Fact Sheet. Rome.

Farrington, J. and P. Deshingkar (2003). Ensuring access to extension for women and vulnerable groups. Paper for the workshop on operationalising reform in agricultural extension in South Africa, New Delhi, 6-8 May 2003. Overseas Development Institute, London.

Ferree, M.M. (1990). Beyond separates spheres: Feminism and family research. Journal of Marriage and the Family, 52: 886-884.

Frijters, P. and B.M.S. van Praag (1995). Estimates of poverty ratios and equivalence scales for Russia and parts of former USSR. Netherlands: Tinbergen Institute Discussion Paper, TI 95-149.

Galasso, E. (1999). Intrahousehold allocation and child labor in Indonesia, Paper presented at the Economist Forum, Washington DC, the World Bank.

Gardiner, J. (ed.) (1997). Gender, care and economics. Basingstoke: Macmillan Press.

Gaye, M. and V. Kelly (1996). L'Utilisation des terres agricoles dans le bassin arachidier du Sénégal: facteurs déterminants des combinaisons culturales et des systèmes de rotations. ISRA/URR, Sud Bassin Arachidier.

Ghuman, S.J. (2003). Women's autonomy and child survival: A comparison of Muslim and non-Muslim in four Asians countries. Demography, 40(3): 419-436.

Goedhart, T., V. Halberstadt, A. Kapteyn and B.M. van Praag (1977). The poverty line: Concepts and measurement. Journal of Human Resources, 12: 503-20.

Gordon, A. and R. Lynne (eds.) (1996). Transforming capitalism and patriarchy: Gender and Development in Africa. Washington: Publishers, Inc.

Greene, W.H. (ed.) (1993) 'Frontier Production Functions', EC-93-20. Stern School of Business, New York University.

Groot, W. and H. Maassen van den Brink (2006). What does education do to our health? Measuring the effect of education on health and civic engagement. Proceedings of the Copenhagen symposium, Paris: OECD.

Grootaert, C.H. and H.A. Patinos (1999). Policy analysis of child labor: A comparative Study. New York, St. Martin's Press.

Gross. R.N. and C. Auffrey (1989). Literacy and health status in developing countries. Annual Review of Public Health, 10: 281-97.

Grown, C., D. Elson and N. Catagay (2000). Special issue on growth, trade, finance, and gender inequality. World Development, 28 (7): 145-55.

Gultierez, L.C. (1998). Infants, illness and mothers time allocation. Mimeo, Center for Policy Research, New York: Syracuse University.

Guyer, J. (1981). Household and community in African studies. African Studies Review, 24(2 and 3): 87-137.

Guyer, J. (1988). Dynamic approaches to domestic budgeting: cases and methods from Africa. In: D. Dwyer and J. Bruce (eds.), A home divided: Women and income in the Third World. Standford: Standford University Press, 155-172.

Haddad, L., J. Hoddinott and H. Alderman (eds.) (1997). Intrahousehold resource allocation in developing countries: Methods, models, and policy. Baltimore, Md., USA: Johns Hopkins University Press for the International Food Policy Research Institute.

Hagenaars, A.J.M. (1986). The perception of poverty. Amsterdam, North-Holland.

Hands, W. (ed.) (2001). Reflections without rules: Economics methodology and contemporary science theory. Cambridge, Cambridge University Press.

Heckman, J. (1976). The common structure of statistical models of truncation, sample selection, and limited dependent variables and a simple estimator for such models. Annals of Economic and Social Measurement, 5: 475-492.

Heckman, J. (1979). Sample Selection bias as a specification error. Econometrica, 47: 153-161.

Hill, S. (ed.) (1994). Managing sickle cell disease in low income families. Philadelphia, PA: Temple University Press.

Hill, S. and M.K. Zimmerman (1995). Valiant girls and vulnerable boys: The impact of gender and race on mothers' care giving for chronically-ill children. Journal of Marriage and the Family, 57: 43-53.

Himmelweit, S. (2001). Decision-making in household. In: Himmelweit, S., Simonetti, R. and Trigg, A. (eds), Microeconomics. Neoclassical and institutionalist perspectives on economic behaviour. London: Thomson Learning and the Open University, pp. 143-177.

Himmelweit, S., R. Simonetti and A. Trigg (eds.) (2001). Microeconomics. Neoclassical and institutionalist perspectives on economic behaviour. London: Thomson Learning and the Open University.

Horney, M. and M.B. McElroy (1988). The household allocation problem: Empirical results from a bargaining problem. Research in Population Economics, 6: 15-38.

Hochschild, A. (ed.) (1991). The second shift. New York: Avon Book.

IFPRI (2000). The key to food security, looking into the household. Washington DC – IFPRI – CGIAR.

Ilahi, N. and S. Jafarey (1999). Market, deforestation and female time allocation in rural Pakistan, Mimeo Montreal: McGill University, Department of Economics.

Ilahi, N. (2000). The Intra-household allocation of time and tasks: What have we learnt from the empirical literature? Development Research Group Poverty Reduction and Economic Management Network. Washington D.C.World Bank.

IMF (International Monetary Fund) (2006). Senegal: Poverty Reduction Strategy Paper – Second Annual Progress Report. Washington, DC.

Jacobson, J. (1992). Gender bias: Roadblock to sustainable development. World Watch Working paper 110, World Watch, Washington DC.

Jacoby, H. (1991). Productivity of men and women and the sexual division of labor in peasant agriculture of the Peruvian Sierra. Journal of Development Economics, 37 (1 and 2): 265-287.

Jacoby, H. (1995). The economics of polygyny in Sub-Saharan Africa: Female productivity and the demand for wives in Cote d'Ivoire. Journal of Political Economy, 103 (5): 938-971.

Jorgenson, D.W. and D.T. Slesnick (1984). Aggregate consumer behavior and the measurement of inequality. Review of Economic Studies, 166: 369-392.

Johannes. P. and J. Jutting (2004). Do community-based health insurance schemes improve poor people's access to health care? Evidence from rural Senegal. World Development, 32 (2): 273-288.

Jones, C. (1983). The mobilization of women's labor for cash crop production: A game theoretic approach. American Journal of Agricultural Economic, 65 (5): 1049-1054.

Jutting, J. (2001). The Impact of community-based health financing on financial protection: A case study in Senegal. World Bank Discussion Paper Series on health, nutrition and population, World Bank, Washington, DC.

Kabeer, N. (1999). the conditions and consequences of choices: reflections on the measurement of women's empowerment. Unites nations research Institute for social development, UNRISD Discussion paper no 108. Geneva.

Kabeer, N. (ed.) (1994). Reversed realities. Gender hierarchies in development thought. London – New York, Verso.

Kabeer, N. (1992). Beyond the threshold: Intrahousehold relations and policy perspectives. Paper prepared for the International Food Policy Research Institute-World Bank. Conference on intrahousehold resource allocation: policy issues and research methods, 12-14 February, Washington DC.

Kanbur, R. (1991). Linear expenditure systems, children as public goods, and intrahousehold inequality. Discussion Paper. England: University of Warwick, Development Economics Research Centre.

Kapteyn, A. (1977). A theory of preference formation. Ph.D. diss., Leyden University.

Kapteyn, A. (1985). Utility and economics. De Economist, 133: 1-20.

Kapteyn, A. and T. Wansbeek (1985). The individual welfare function. Journal of Economic Psychology, 6: 333-363.

Kapteyn, A., P. Kooreman and R. Willemse (1988). Some methodological issues in the implementation of subjective poverty definitions. The Journal of Human Resources, 23:222-242.

Kelly, R. and S. Shortall (2002). Farmers wives': Women who are off-farm breadwinners and the implications for on-farm gender relations. Journal of Sociology, 38 (4): 327-343.

Kelly, V., B. Diagana, T. Reardon, M. Gaye and E. Crawford (1996). Cash crop and food grain productivity in Senegal: Historical view, new survey evidence, and policy implication. MSU International Development Paper No.20, East Lansing, Michigan.

Kevane, M. (1998). Extrahousehold norms and intra-household bargaining: Gender in Sudan and Burkina Faso. Leavey School of Business - Economics Department. Santa Clara University.

Kevane, M. and B. Wydick (1999). Social norms and the allocation of women's labour in Burkina Faso. Review of Development Economics, 5 (1): 119-129.

Khandker, S.R. (1988). Determinants of women's time allocation in rural Bengladesh. Economic Development and Cultural Change, 37 (1): 111-126.

Kimhi, A. (2000). Time allocation and off farm activities (is part-time farming really a step in the way out of agriculture). American Journal of Agricultural Economics, 82 (1): 38-48.

Kong, K.M., H. Lee and J. Lee (2005). Health uncertainty and precautionary saving: Evidence from Korea. SCID working paper 262. Stanford Institute for Economic Policy Research. Stanford University.

Lado, C. (1992). Female labour participation in agriculture production and the implications for nutrition and health in rural Africa. Social Science and Medicine, 34 (2): 787-807.

Lattell, R.B. (1966). The Scree test for the number of factors. Multivariate Behavioral Research, 1: 215-76.

Lee, E. and M. Loufti (1992). La stagnation économique et l'emploi au Sénégal. In: Ajustement, emploi et développement au Sénégal. Genève: BIT, 1-21.

Lewis, M. and M. Lockhed (2008). Social exclusion and the gender gap in education. Policy research working paper, no. WPS 4562, 2008. Washington: World Bank, Human Development Network.

Lifanda, K.N. (2005). Education and the training of women and the girld-child. BAS/PE 2005/ME. Online discussion for the review of Beijing + 10 (UNESCO/UNICEF). January 10-04 February 2005.

Lilja, N., J.H. Sanders, C.A. Durham, H. de Groote and I. Dembele (1996). Factors influencing the payments to women in Malian agriculture. American Journal of Agricultural Economics, 78: 1340-1345.

Lilja, N., T.F. Randolph and A. Diallo (1998). Estimating gender differences in agricultural productivity: Biases to omission of gender-influenced variables and endogeneity of regressors. Paper presented at the American Agricultural Economics Association annual meeting in Salt Lake City, Utah, 2-5 August, 1998.

Lim Sung, S. A. Winter-Nelson and M. Arends-Kuenning (2007). Household bargaining power and agricultural supply response: Evidence from Ethiopian coffee growers. World Development, 35 (7): 1204-1220.

Locoh, T. (1996). Changements de rôles masculins et féminins dans la crise: La révolution silencieuse. In: Coussy, J. and Vallin, J.(eds.), Crise et population en Afrique. Crises économiques, programmes d'ajustement et dynamiques démographiques. Paris: Les Études du CEPED n°13, 445-470.

Lopez-Claros, A. and S. Zahidi (2005). Women's empowerment: Measuring the global gender gap. Geneva: World Economic Forum. From http://www.weforum.org/pdf/global_Competitiveness_Reports/Reports/gender_gap.pdf (accessed 5 August 2005).

Lucas, R. and S.B. Odded (1993). Making a miracle. Econometrica, 61(2):251-272.

Lukmanji, Z. (1992). Women's workload and its impact on their health and nutritional status. Progress in Food and. Nutrition Sciences, 16(2):163-79.

Lundberg, M. (2003). Do savings predict death? Precautionary savings during an epidemic. Washington D.C: Mimeo. The World Bank.

Lundberg, S. (2005). Gender and household decision-making. Lectures Notes University of Siena International School of Economic Research. XVIII Workshop on Gender and Economics. 4-7th July 2005, Certosa di Pontignano.

Lundberg, S. and R.A. Pollak (1993). Separate spheres bargaining and the marriage market. Journal of Political Economy, 101 (6): 988-1010.

Lundberg, S. and R.A. Pollak (1994). Non-cooperative models of marriage. American Economic Review, 34 (2):132-137.

Lundberg, S. and R. Pollak (1996). Bargaining and distribution in marriage. The Journal of Economic Perspectives, 10 (4): 139-158.

Lundberg, S, R.A. Pollak and T.J. Wales (1997). Do husbands and wives pool their resources? Evidence from the United Kingdom child benefit. Journal of Human Resources, 32 (3): 463-480.

Maasoumi, E. (1989). Composite indices of income and other developmental indicators: A general approach. Research on Economic Inequality, 1: 269-286.

Mackintosh, M. (ed.) (1989). Gender, Class and Rural Transition. Agri business and the food crisis in Senegal. New Jersey, USA: Zed Books UK and Atlantic Highlands 07716.

Maertens, M., L. Colen and J.F.M. Swinnen (2008). Globalization and poverty in Senegal: A worst case Scenario? Katholieke Universiteit Leuven, Licos discussion paper No. 217.

Manser, M. and M. Brown (1980). Marriage and household decision-making: A bargaining analysis. International Economic Review, 21(1): 31-44.

Maurique, J. and H. Jensen (2008). Working women and expenditures on food away from home and at home in Spain. Journal of Agricultural Economics, 49 (3): 321-333.

Mayoux, L. (2001). Trackling the dark side: social capital, women's empowerment and microfinance in Cameroun. Development and Change, 3(3): 445-464.

Mbilinyi, M. (1988). The invention of female farming system in Africa: Structural adjustment in Tanzania. Workshop on economic crisis, household strategies and women's work, Ithaca, New York: Cornell University.

Mbow, P. (2009). Evolving role for Senegalese women in Religion. Common ground news 2009. From http://www.commongroundnews.org/.

McElroy, M.B. (1990). The empirical content of Nash-bargained household behavior. Journal of Human Resources, 25 (4): 559-583.

McElroy, M.B. and M.J. Horney (1981). Nash-bargained household decisions: Toward a generalization of the theory of demand. International Economic Review, 22 (2): 333-349.

Mcpake, B., L. Kumaranayake and C. Normand (eds.) (2008). Second edition. Health economics. An international perspective. London: Routledge.

Meherun, A. (2006). Intra-household bargaining and investment in child health. Department of Economics, Carleton College, Northfield, Minnesota. D12, 121, 012.

Mehta, M. (1998). Our lives are not different from that of the buffaloes. In: D. Rocheleau, T.B. Slayte and E. Wangari (eds.), Feminist political ecology. London and New York: Routledge, 1181-207.

Moghadam, V.M. (2004). Patriarchy in transition: Women and the changing family in Middle East. Journal of Comparative Family Studies, 7 (2): 137-163.

Momsen, J.H. (ed.) (2004). Gender and development. London and New York: Routledge Perspectives on Development, Series Editor: Binns, T., University of Sussex.

Morris, L. (ed.) (1990). The workings of the household. Cambridge: Policy Press.

Moser, C.O.N. (ed.) (1993). Gender planning and development: Theory, practice and training. London, New York: Routledge.

Mtshali, S.M. (2002). Household livelihood security in rural KwaZulu-Natal, South Africa, PhD Thesis. Wageningen University.

Müller, T.R. (2005). HIV/AIDS, Gender and rural livelihoods in sub-Saharan Africa. AWLAE Series No.2. Wageningen: Wageningen Academic Publishers.

Müller, T.R. (2004). HIV/AIDS and Agriculture in sub-Saharan Africa. AWLAE Series No.1. Wageningen: Wageningen Academic Publishers.

Mullins, G., L. Wahome, P. Tsangari and L. Maarse (1996). Impact of intensive dairy production on smallholder farm women in Coastal Kenya. Human Ecology, 24: 231-253.

Muriuki, H.G., M.O. Owango and S. Staal (1997). Dairy co-operative and policy reform in Kenya: Effects of livestock service and milk market liberalisation. FAO, Rome (Italy). Market oriented dairying – the role of cooperatives and NGOs. Proceedings of a seminar.

Mutangadura, G. (2000). Household welfare impacts of mortality of adults females in Zimbabwe: Implications for policy and programme development. Paper presented at the AIDS and Economics Symposium, Durban, 7-8 July 2000.

Myres, D. (1999). Close relationship and quality of life. In: Kahneman, D., Diener, E. and Schwartz, N. (eds.), Well-being: The Foundation of Hedonic Psychology. New York, Chapter 19: Russell Sage Foundation.

Nash, J. (1950). The bargaining problem. Econometrica, 18(2): 155-162.

Nash, J. (1953). Two-person bargaining games. Econometrica, 21(1): 128-140.

Ndiaye (2001). Endettement extérieur, PAS et pauvreté au Sénégal from: http://www.france.attac. org/.

Ndiaye, B. (2005). Les déterminants de la pauvreté en milieu rural au Sénégal. Document de travail IDEP, Dakar/Sénégal.

Ndiaye, D.A. (2007). Les aspects socioculturels de la négociation intra ménage en zone rurale du Sénégal. Rapport de la Coopération Senegalo-Suisse Dakar.

Ndirangu, L. (2007). Household's vulnerability and responses to shocks: Evidence from rural Kenya. Ph.D. Dissertation, Wageningen University. 155 p.

Newell, S. (1993). The Superwoman syndrome. Gender difference in attitudes towards equal opportunities at work and towards domestic responsibilities at home. Work, Employment and Society, 7 (2): 275-89.

Ng, Y.-K. (1997). A case of happiness, cardinalism, and interpersonal comparability. Economic Journal, 107: 1848-58.

Ngom, P., S. Wawire, T. Gandaho, P. Klissou, T. Adjimon, M. Seye, E. Akouanou and L. Winter (2000). Intra-household decision-making on health and resource allocation in Borgou, Benin. Working paper, HNR-A-00-98.

Niehof, A. (2003). Gendered dynamics of food security. In: Ulmer K. (ed.), No security with-out food security – No food security without gender equality. Brussels: APRODEV, pp. 59-66.

Niehof, A. (2004). A micro-ecological approach to home care for AIDS Patients. Medische Antropologie Tijdschrift over gezondheid en cultuur, 16 (2): 245-265.

Niehof, A. and L.L. Price (2001). Rural livelihood system: A conceptual framework. UPWARD Working Paper Series No. 5. Wageningen: WU-UPWARD.

Nowak, B.S. (2008). Environmental degradation and its gendered impact on Coastal livelihoods: options among Btsisi' households of Peninsular Malaysia. Development, 51 (2): 186-192.

Nussbaum, M. and A. Sen (eds.) (1993). The quality of life, Clarendon Press, Oxford.

OECD (2001). Measuring productivity. OECD manual measurement of aggregate and industry-level productivity growth. Paris, OECD.

O'Neill, J. and D. O'Neill (2005). What do wage differentials tell about labour market discrimination? Working Paper no 11240. Cambridge, MA, National Bureau of Economic Research.

Oswald, A. (1997). Happiness and eonomic peformance. The Economic Journal, 107: 1815-1831.

Overbeek, G., S. Efstratoglou, M.S. Haugen and E. Saraceno (1998). Labour situation and strategies of farm women in diversified rural areas of Europe. Research proposal CT94 2414 for AIR-programme, Office for Official Publication of the European Communities, Luxembourg.

Oxaal, Z. and S. Cook (1998). Health and poverty: A gender analysis. Bridge Report no. 46, prepared for the Swedish International Development Agency. IDS (Institute of development studies), Brighton.

Pahl, J. (1983). The Allocation of money and the structuring of inequality within marriage. Sociological Review, 31 (2): 237-262.

Parducci, A. (ed.) (1995). Happiness, pleasure, and judgment: The contextual theory and its applications. Mahwah, New Jersey Hove, UK: Lawrence Erlbaum Associates, Publishers.

Phipps, S.A. and S. Burton (1998). What's mine is Yours? The influence of male and female incomes on patterns of household expenditures. Economica, 65 (260): 599-613.

Pilon, M., M.M. Seidou and C. Tichit (1997). Les femmes chefs de ménage: aperçu général. In: Pilon, M., Locoh., V. and Vimard, P. (eds.), Ménages et familles en Afrique: Approches des dynamiques contemporaines. Paris: Etudes du CEPED no15, 168-191.

Plug, E.J.S. (1997). Leyden welfare and beyond. Thesis Publishers and the Tinbergen Institute Series, no. 148, Amsterdam.

Plug, E.J.S. and B.M.S. van Praag (1995). Family equivalence scales within a narrow and broad welfare context. Journal of Income Distribution, 4(2): 171-186.

Plug, E.J.S. and B.M.S van Praag (1995). Welfare and Intelligence, Timbergen Discussion paper TI 1-95-196. University of Amsterdam.

PNUD (1998). Rapport mondial sur le développement humain, Paris, 254 p.

Pollak, R. (2005). Bargaining power in marriage: Earnings, wage rates and household production. National Bureau of Economic Research. Cambridge, MA 02138. Working paper 11239. From http:www.nber.org/papers/w11239.

Pyke, K. (1996). Class-based masculinities: The interdependence of gender, class, and interpersonal power. Gender and Society, 10 (5): 527-549.

Pourcet, G. (1992). La ville, l'informel et l'environnement. Afrique contemporaine, 161: 118-187.

Potuchek, J.L. (1992). Employed wives' orientations to breadwinning: A gender theory analysis. Journal of Marriage and the Family, 54: 548-558.

Quisumbing, A.R. (1994). Intergenerational transfers in Philippine rice villages: Gender differences in traditional inheritances customs. Journal of Development Economics, 43(2): 167-195.

Quisumbing, A. (1996). Male-female differences in agricultural productivity: Methodological issues and empirical evidence. World Development, 24(10): 1579-1595.

Quisumbing, A. (2007). Gender differences in agricultural productivity: A survey of empirical evidence. FCND discussion paper no. 5. From: http://en.scientificcommons.org/42820001.

Quisumbing, A. and B. de la Briere (1999). Women's Assets and intrahousehold allocation in Bangladesh: Testing measures of bargaining power. FCND Discussion Paper 86, IFPRI, Washington, DC.

Quisumbing, A. and J. Maluccio (2000). Intrahousehold allocation and gender relations: New empirical evidence from four developing countries. FCND Discussion Paper 84, IFPRI, Washington, DC.

Rainwater, L. (ed.) (1974). What money buys: Inequality and the social meaning of income. New York: Basics Books.

Ravallion, M. (ed.) (1994). Poverty comparisons. Fundamentals in pure and applied economics, 56, Chur, Switzerland: Harwood Academic Press.

Reardon, T., C. Delgado and P. Matlon (1992). Determinants and effects of income diversification amongst farm household in Burkina Faso. Journal of Development Studies, 28 (2): 264-296.

Robbins, L. (ed.) (1932). An essay of the nature and significance of economic science. London, MacMillan.

Rogers, S. and D. Deboer (2001). Change in wives' income: Effects on marital Happiness, psychological well-being and the risk of divorce. Journal of Marriage and Family, 63 (2): 458-472.

Rosenzweig, M. and O. Stark (1989). Consumption smoothing, migration and marriage: Evidence from Rural India. Journal of Political Economy, 97: 905-926.

Ross, C.E. and C. Wu (1995). The links between education and health. American Sociological Review, 60: 719-745.

Rouis, M. (1994). Senegal, stabilization, partial adjustment and stagnation. In: Hussain, I. and Faruqee, R. (eds.), Adjustment in Africa: Lesson from country case studies. Washington, DC: The World Bank, 286-351.

Rowling, M.R. (2008). Women farmers toil to expand Africa's food supply. Global Sisterhood Network. From http://www.global-sisterhood-network.org/content/view/2244/59/.

Rugalema, G. (1999). Adult mortality as entitlement failure. AIDS and the crisis of rural livelihoods in a Tanzania village. PhD thesis Institute of Social Studies, The Hague, 227 pp.

Saito, K., H. Makonnen and D. Spurling (1994). Raising the productivity of women farmers in Sub-Sahara Africa. Discussion Paper 230. Washington DC: World Bank.

Samuelson. P. (1956). Social indifference curves. Quarterly Journal of Economics, 70 (1): 1-22.

Saul, M. (1989). Separateness and Relation: Autonomous income and negociation among rural Bobo women. In: Wilk, R. (ed.), The household economy: Reconsidering the domestic mode of production. Boulder: Westview Press, 171-197.

Schultz, T.P. (1990). Testing the neoclassical model of family labor supply and fertility. Journal of Human Resources, 25 (4): 599-634.

Schwarz, N. and F. Strack (1991). Evaluating one's life: A judgment model of subjective well-being. In: F. Strack, M. Argyle and N. Schwarz (eds.), Subjective well-being. An interdisciplinary perspective. Oxford: Pergamon Press, 27-47.

Schyns, P. (2001). Income and satisfaction in post communist Russia. Journal of Happiness Studies, 2(2):173-204.

Seck, T.A. (ed.) (1997). La Banque mondiale et l'Afrique de l'Ouest. L'exemple du Sénégal. Publisud, Paris.

Sen, A.K. (ed.) (1987). The standard of living. Cambridge University Press, The Tanner lectures.

Sen, A.K. (ed.) (1984). Resources, values and development. Cambridge, Mass., Harvard University Press.

Sen, A.K. (1990). Gender and cooperative conflicts. In: Tinker, I. (ed.), Persistent inequalities. Oxford: Oxford University Press, 123-49.

Sénégal, Ministère de l'Economie et des Finances (MEF), DPS. (1997). Enquêtes Sénégalaise auprès des Ménages (ESAM), 1994-1995, 178 p.

Sénégal, CSE. (2003). L'utilisation des terres agricoles au Sénégal. Rapport final: Projet FAO/CSE (Centre de suivi Ecologique). Sénégal, Dakar, 51 p.

Sénégal, Ministère de l'Agriculture (DA). (2004). Evaluation de l'application des engagements du Sommet Mondial sur l'Alimentation (SMA) de 1996 a Rome. Sénégal: rapport final mai 2004, 62 p.

Sénégal, Ministère de l'Economie et des Finances (MEF, ANSD) (2005). DPS, Enquêtes démographiques et de santé II (EDS II), 1992-1993, 284 p.

Sénégal, Ministère de l'Economie et des Finances (MEF, ANSD) (2005). DPS, Enquêtes démographiques et de santé III (EDS III), 1995-1997, 238 p.

Sénégal, Ministère de l'Economie et des Finances (MEF, ANSD) (2007). Enquêtes de suivi de la pauvreté au Sénégal ESPS, 2005-2006. Rapport National 2007, 89 p.

Sénégal, FONGs (2008). La filiere riz au Senegal. Rapport de travail, Senegal: SOS Faim 2008.

Sevilla-Sanz, A. (2005). Social effects, household time allocation, and the decline in union formation. working paper, Congressional Budget Office.

Shortall, S. (1992). Power analysis and farm wives: An empirical study of the power relationship affecting women on Irish farms. Sociologia Ruralis, 32 (4): 431-451.

Singh, I., L. Squire and J. Strauss (eds.) (1986). Agricultural household models:Extensions, applications and policy. Baltimore: John Hopkins University Press.

Smith, J. (1999). Healthy bodies and thick wallets: The dual relation between health and economic status. The Journal of Economic Perspectives, 13 (2): 145-166.

Smith, L.C. and J.P. Chavas (1997). Commercialization and the balance of women's dual roles in non-income-pooling West African Household. American Journal of Agricultural Economics, 79 (2): 589-94.

Smith, L.C. and J.P. Chavas (1999). Supply response of West African agricultural household: Implication of intra-household preference heterogeneity. Food Consumption and Nutrition Division. IFPRI, Washington, DC.

Staal, S.J. (2002). The competitiveness of smallholder dairy production: Evidence from sub-Sahara Africa, Asia and Latin America. In: D. Ragnekar and W. Thorpe (eds.), Smallholder dairy production and marketing opportunities and constraints:Proceedings of a south-south workshop, Anand, Gujarat (India). Australian Center for International Agricultural Research. NDDB, 250-264.

Stokey, N.L. (1991). Human capital, product quality, and growth. Quaterly Journal of Economics, 106 (2): 587-616.

Strauss, J. and D. Thomas (1995). Human resources: Empirical modeling of household and family decisions. In: J. Behrman and T.N. Srinivasan (eds.), Handbook of development economics, Vol. IIIA. Amsterdam: Elsevier, 1883-2023.

Suwadu, M., J. Sakho and M. Benoit-Cattin (2007). Cereals supplies in rural families of the Senegalese Groundnut Basin. Who is responsible for meeting family food needs? Paper prepared for presentation at the 106th seminar of the EAAE: Pro-poor development in low income countries: food, agriculture, trade and Environment, 25-27 October 2007. Montpellier, France.

Tangka F., E.A. Ouma, S.J. Staal and B. Shapiro (1999). Women and the sustainable development of market-oriented dairying: Evidence from the highlands of East Africa. Paper presented at the International Sustainable Development Research Conference held at University of Leeds, Leeds, UK, 25-26 March 1999.

Tapouzis, D. (ed.) (2000). Measuring the impact of HIV/AIDS on the agricultural sector in Africa. Geneva, UNAIDS Best Practice Collection.

Thompson, L. (1993). Conceptualizing gender in marriage: The case of marital care. Journal of Marriage and the Family, 55 (3): 557-569.

Tipping, G. and M. Segall (1995). Health care seeking behaviour in developing countries. Brighton, Institute of Development Studies: IDS Development Bibliography no 12.

Treas, J. (1993). Money in the Bank: Transaction costs and the economic organization of marriage. American Sociological Review, 58 (5): 723-734.

Udry, C.H., J. Hoddinott, H. Alderman and L. Haddad (1995). Gender differentials in farm productivity: Implications for household efficiency and agricultural policy. Food Policy, 20 (5): 407-423.

Udry, C.H. (1996). Gender, agricultural production, and the theory of the household. Journal of Political Economy, 104 (5): 1010-1046.

UNAIDS/WHO (2003). AIDS epidemic update December 2003, Geneva: UNAIDS/WHO.

UNAIDS/WHO (2006). Report on the global AIDS epidemic: Executive summary, a UNAIDS 10th anniversary special edition. Geneva: from: http://www.unaids.org/en/HIV_data/2006GlobalReport/default.asp.

UNAIDS/WHO (2008). AIDS outlook. UNAIDS World AIDS day report, Geneva: UNAIDS/WHO.

UNDP (2008). Empowered and equal: Gender equality strategy 2008-2011. UNDP report, Bureau for Development Policy, World Bank, Washington DC.

Van Praag, B.M.S. (1968). Individual welfare functions and consumer behavior: A theory of rational irrationality. Ph.D. diss. University of Amsterdam.

Van Praag, B.M.S. (1971). The welfare function of income in Belgium: An empirical investigation. European Economic Review, 2 (3): 337-69.

Van Praag, B.M.S. (1977). The perception of welfare inequality. European Economic Review, 10(2): 189-207.

Van Praag, B.M.S., A. Hagenaars and J. van Weeren (1982a). Poverty in Europe. Review of Income and Wealth, 28 (3): 345-59.

Van Praag, B.M.S., J.S. Spit and H. van de Stadt (1982b). A comparison between the food ratio poverty line and the Leyden poverty line. The Review of Economics and Statistics, 64 (4): 691-694.

Van Praag, B.M.S. and J. van Weeren (1983). Some panel-data evidence on the time-discounting mechanism in the formation of value judgments on income with application to social security and income policy. Report 83.22. Center for Research in Public economics, Leyden University.

Van Praag, B.M.S. and N.L. van der Sar (1988). Household cost functions and equivalence scales. Journal of Human Resources, 23: 193-210.

Van Praag, B.M.S. and E.J.S. Plug (1995). New Developments in the measurement of welfare and well-being. The Ragnar Frisch Centeniali, Tinbergen Institute Discussion paper TI 1995-60, the Netherlands.

Van Praag, B.M.S. and P. Frijters (1999). The measurement of welfare and well-being: The Leyden approach. In: D. Kahneman, E. Diener and N. Schwarz (eds.), Well-being: the foundations of hedonic psychology. Russell Sage Foundation, New York, 413-433.

Van Praag, B.M.S. and B.E. Baarsma (2000). The shadow price of aircraft nuisance. Tinbergen Institute Discussions Papers, 00-004/3, the Netherlands.

Van Praag, B.M.S. and A. Ferrer-i-Carbonell (eds.) (2004). Happiness quantified: A satisfaction calculus approach. Oxford: Oxford University Press.

Varadharajan, S. (2003). The pitfalls of bargaining power in intra-household analysis. Mimeo, Department of Economics, Cornell University.

Veenhoven, R. (1991). Questions on happiness: classical topics, modern answers, blind spots. In: Strack, F., Argyle and Schwarz, N. (Eds.), Subjective well-being. An interdisciplinary perspective. Oxford, Pergamon Press.

Veenhoven, R. (1996). Happy life expectancy: A comprehensive measure of quality of life in nations. Social Indicators Research, 39(1): 1-58.

Vermeulen, F. (2002). Collective household models: principles and main results. Journal of Economic Surveys, 16 (4): 533-564.

Venema, B. (1986). The changing role of women in Sahelian Agriculture. In: L. Creevey (ed.), Women farmers in Africa: Rural development in Mali and the Sahel. Syracuse: Syracuse University Press, 249-380.

Verwimp, P. (1999). Elasticity returns to education in off-farm activities in rural Ethiopia: Center for Economic Studies. Discussions paper CES 9903. Katholieke Universiteit Leuven.

Visvanathan, N., L. Duggan, L. Nisonoff and N. Wiegersma (eds.) (1997). The women, gender & development reader. Zed Books Ltd, London and New Jersey.

Walton, E. and D. Tacheuchi (2005). Contextualizing the education and health status association: Evidence from a national study on Asian American, Paper presented at the annual meeting of American Sociological Association. Philadelphia, PA, USA.

Wansbeek, T. and A. Kapteyn (1983). Tackling hard questions by means of soft methods: the use of individual welfare function in socio-economic policy. Kyklos 36: 249-269.

Ward-Batts, J. (2003). Out of the wallet and into the purse: Using micro data to test income pooling, Claremont College Working paper 2003-10.

Ware, A.T. (1979). Wolof farmers and Fulani herders: A case study of drought adaptation in the diourbel region of Senegal. Ph.D. Dissertation. Michigan.

Warner, J.M. and D.A. Campbell (2000). Supply response in an agrarian economy with non-symmetric gender relations. World Development, 28 (7): 1327-1340.

Whatmore, S. (ed.) (1991). Farming women: Gender, work and family enterprise. London: Mcmillan.

Wheelock, J. and E. Oughton (1996). The household as a focus for research. Journal of Economic Issues, 30 (1): 143-59.

Whitehead, A. (1990). Rural women and food production in Sub-Saharan Africa. Ch. 11 in: J. Drèze and A. Sen (eds.), The political economy of hunger Entitlement and Well-Being (Vol. 1): Oxford: Clardendon Press, 425-473.

WHO (1995). The World Health Report. World Health Organization. Geneva.

WHO (2003). The world health report 2003. Shaping the future: Chapter I, Global health: Today's challenges. World Health Organization. Geneva.

WHO (2006a). Working together for health. World Health Organization. Geneva.

WHO (2006b). Gender, women and health: draft strategy. WHO. Executive Board, EB, 120/6. Geneva.

Winters, P. and V. Chiodi (2008). Human capital investment and long-term poverty reduction in rural Mexico. Working paper no. 2008-51. Paris School of Economics- LEA-INRA.

Woolley, F. (2003). Control over money in marriage. In: S. Grossbard-Shechtman (ed.), Marriage and the economy: Theory and evidence from advanced industrial societies. Cambridge: Cambridge University Press, 105-128.

World Bank (1993). The benefits of education for women. Human Resources Development and Operation Policy. Dissemination Notes No.2 March. Washington DC.

World Bank (1997) Expanding the measure of wealth, indicators of environmentally sustainable development, Washington, DC, World Bank.

World Bank (1999). Nicaragua poverty assessment. The World Bank, Washington DC.

World Bank (2001). Engendering development, World Bank Policy Report. Washington DC.

World Bank (2004). Private sector for Models for Poverty Reduction. Report on the Field visit to Sichuan and Zhejiang Provinces of China. IFC/ World Bank. Washington DC.

World Bank (2004). World development indicators. Washington DC.

World Bank (2005). Gender and shared growth in Sub-Saharan Africa. Briefing notes on critical gender issues in Sub-Saharan Africa. World Bank report. Washington, DC.

Xinyu, C. and C. Yanwei (2007). Gender role based differences in time allocation: Case study of Shenzhen, China. Transportation Research Board. Annual meeting 2007. paper 07-0574. 23 p.

Zimmerman, M.K. and S.A. Hill (1999). Health care as a gendered system. In: Saltzman Chafetz, J. (ed.), Handbook of the sociology of gender. New York: Kluwer Academic/Plenum Publishers, 483-518.

Summary

In Senegal, one of the ten poorest countries in the World, the national Strategic Document for Poverty Reduction (SDPR) and other strategic documents (from development partners and NGOs) emphasize a gender approach to reducing the current level of poverty of more than 60% of rural households. In addition to food insecurity, people's health problems have also been considered a serious cause of livelihood uncertainty in rural households. These health problems have recently been exacerbated by the spread of HIV/AIDS causing a serious disaster in many African countries. Because of the interdependence of partners (husbands and wives) in the farming system, and the importance of individual benefits in reaching adequate levels of well-being of the household members, the separate roles and obligations of both partners in the household need to be better understood. The identification of the separate roles of individuals in household-decision-making on the allocation of resources will contribute to the understanding of intrahousehold dynamics and, consequently, to designing more suitable policy programs.

The aim of this study is to make a contribution to a better understanding of the intrahousehold dynamics between husband and wife with respect to decision-making. A better understanding of intrahousehold resource allocation also implies an understanding of the effect of time allocation and labour supply, in particular on individual and household income. The daily provision of household goods in rural Senegalese households is seen as a result of partners' earnings and decision-making concerning the provision of household and private goods. These issues are often misunderstood and disregarded in developing countries. This thesis provides empirical insights into the way in which partners in rural Senegalese households behave with respect to resource allocation, income procurement and health problems, paying attention to the major constraints they are confronted with. Furthermore, welfare and well-being have been measured by using subjective methods never before applied in West-African countries. Several research questions have been specified and will be considered in summarizing the results of the empirical chapters of this study.

The study is based on cross-sectional data collection in 2006-2007 among 300 households in two different types of farming areas. One is the agro-pastoral (AP) area and the other the sylvo-pastoral (SP) area. The two research areas differ in physical and natural resources and socio-economic characteristics, resulting in different livelihood levels and procurement opportunities.

In Chapter 3, our first empirical chapter, we investigate whether female and child-labour inputs positively contribute to household earnings, and whether female time allocation is affected by farm- and housework. Secondly, this chapter provides insight into the extent to which earnings from farming of partners depend on individual, household and socio-cultural characteristics and varies across areas. In doing so, we tried to answer the following three research questions: (1) How does the distribution of intrahousehold activities and responsibilities in farming differ across SP and AP areas? (2) What are the structural and socio-economic determinants that are associated with the gender division of resource allocation and control over income across different systems of production? (3) How does the gender allocation of time within the household may be correlated with the earnings from of agricultural activities?

Since some 10% of the sample did not earn any income from livestock and 49% did not earn income from crops, and so as to avoid selection bias, we estimated the regressions analysis with respect to time allocation and earnings using Heckman's two-step model. The time allocated to livestock and crop production was negatively dependent on time allocated to housework, especially for wives and girls. Although women spent about 6 hours on crops compare to 8 hours for men (almost the same), they received less income (working in crops, men earn almost 34 times as much as women). These results suggest that in crop work male household heads are by far the main earners. For livestock work women received 18% of total earnings compared to 76% for men. However, females were more engaged in milk production where their earnings were 4.5% (of total earnings) compared to 1% for men. Overall from crops and livestock, men earn five times more than women do In fact, in Senegalese farming, women and girls are responsible for milking and marketing milk at the nearest markets. In addition to farming activities, earnings came from off-farm activities and cash transfers from migrants representing 5 and 3%, respectively, of total household earnings. The sylvo-pastoral (SP) area gains more from livestock, while the agro-pastoral (AP) area is specialized in crops. However, SP farmers are better off than AP farmers. For lack of land in the AP area, livestock and crop production have to compete for it.

Livestock production and crop production are each other's substitutes and compete in labour force participation and time use. Keeping more livestock, especially cattle, showed a negative association effect on the participation in crop production. The participation of females in terms of labour-time allocation positively increases income both from livestock and crops, while for male heads there was a positive association only with livestock. Thus the productivity of agricultural activities is gendered and also highly dependent on the time women and girls spend on housework, and the time boys and other members spend on crops and livestock. The husband's caring for ill household members negatively affected household income from livestock. For females, caring had a positive effect on livestock income, but overall, their involvement in housework negatively affected household and individual earnings. Regarding the determinants of time allocation, women are constrained by the time which they have to spend on housework, and by the time which their husbands spend on livestock and crops. It appeared that women are forced to supply more labour to farming when their husbands work more hours either in livestock or in crop production. An increase in the burden of housework is also negatively correlated with females' time allocation to farming, hence forced them to be more efficient in order to produce the same output.

Household earnings are also positively associated with off-farm activities that allow for more investments in inputs and machinery. A higher level of men's education is positively correlated with their income from livestock, while for women, the same effect was visible for crop production. However, better-educated wives are less interested in participating in farming. Looking at socio-economic factors that may affect household and women's earnings, we found that access to land had a positive correlation with earnings from livestock and crops. In addition wives' savings allowed women to be more productive, especially in livestock. However, savings decreased her probability of participating in crop production. Access to credit allowed women to be less interested in farming and to be more oriented towards diversification, i.e. engaging in other activities than farming (trade and small entrepreneurship). Cash transfers from migrants (non-labour income) had a negative correlation with farming, especially on the livestock section.

The distance of most rural areas to markets and big centres disfavoured wives' and husbands' incomes from livestock and crops. With respect to the effect of health status on productivity, the results showed that a good health status affected especially the husband's earnings from crop production.

Chapter 4 tries to answer two important research questions: How do men and women engage in food and health provision in terms of decision-making and how can we measure women's power? Empowerment was indicated by the components of a factor analysis of items capturing women's power, in order to analyze women's status in the household. The findings indicate that, although male heads mainly decide on goods provision, in both areas women too were responsible for providing food and managing finance. A comparison between the two areas showed that SP women have more voice in decision-making concerning the purchase of food and household finance management than AP women. With respect to health issues, while men are the main decision makers in household finance, females are the main care givers for ill members in the household. The results suggest that health costs were not met through consensual decision-making or pooling of the spouses' incomes, as in the case of food provision, but were covered by money mostly controlled by the husband. Alternative decisions on health expenditures were related to the recourse to traditional healers (not always implying cash payments). In the event of illness this was observed for 14% of households. A comparison between the areas shows that jointly with teir husbands, SP women made more decisions on health cost and health care (27%) than AP women (14%). SP women were more responsible for health care expenditures than AP women (12 vs. 5%). These results reinforce the evidence of the wife's power in the SP area where women are economically more comfortable.

The factor analysis resulted in four factors indicating the relative status of women: (1) permission to move freely outside the household, (2) voice in food provision, (3) income management and (4) the spouses' mutual willingness to share their income. OLS regression showed that the factors were positively affected by the age of women, their financial contribution to food provision, their non-labour income (cash transfers from migrants) and women's assets (livestock endowments). The findings suggested that experience, assets and capital endowments gave women a more powerful voice in decision-making and in increasing the household's well-being. Moreover, their assets and capital endowment made women feel less constrained in going to the market and to the hospital. The constraint of having to ask permission varied across zones and was more visible in the AP area.

While the results in Chapter 3 described earnings and those in Chapter 4 highlighted the characteristics of decision-making on food and health provision, Chapter 5 extended the discussions by giving more insight into the destination of the earnings.

In Chapter 5, we describe expenditure patterns and behavioural attitudes by answering the following question: how can we capture the bargaining relationship between husband and wife and its effects on expenditures?

The results showed that both husband and wife contribute to household expenditures on food, schooling, health and clothes. Women participated in all expenditures with a share of 20% on food and of 21% on other goods (schooling, health and clothes). Looking into the way expenditures were made, the results showed overall rejection of the income-pooling hypothesis. However, there is a predominance of pooling in the AP area where only for schooling expenditures the income was not pooled. The hypothesis that one additional franc of the wife's income is spent in the same way as one additional franc of the husband's income, was rejected for expenditures related to health, clothing, tea and coffee. For food and schooling, it seems that the partners spend income in a consensual way, leading to a kind of income sharing between husband and wife. Also, our results provide evidence that husband's and wife's income is not spent on the same goods. Area appears to matter as well: an increase in the wife's income in the SP area increases food, health and schooling expenditures, whereas in the AP area, it does not.

Household size, ethnicity and religion further affect expenditures. While the results concerning household size show that it raises expenditures for almost all goods, area and ethnicity show differences in consumption priorities. AP farmers are less oriented towards private expenditures. Religion and education also matter, since being Catholic or Muslim affected the way of spending on goods, and a higher level of the husband's education induces more expenditure on child education and school attainment.

The non-cooperative model highlights that men gain higher utility in their private expenditures than women. However, both husband and wife are more sensitive to the consumption of household goods (food, schooling, health) than to the consumption of private goods (clothing and other). Monetary compensation from husbands to wives for their contribution to farming, is negatively associated with wife's time allocated to livestock (which is determinant of her earnings) and the SP area, and is positively correlated with the wife's responsibility in managing finance. Furthermore, an important exogenous factor that affects the equilibrium values of the parameter (sensitivity of utility to private expenditures) is the ethnicity which reflects difference between community's characteristics.

Chapter 6 goes beyond the individual estimation of the elasticity of utility of private-goods consumption, and uses subjective methods to capture the welfare and well-being of husband and wife separately. In Chapter 6 we have answered the following question: how can we measure the welfare and well-being of husbands and wives? In doing so, we used the Leyden approach. The hypothesis related to husbands' and wives' answers to the minimum income question (MIQ) and the income evaluation question (IEQ) which they answered separately.

The results show that the individual welfare function of income (WFI) based on the IEQ is different for husband and wife. The WFI showed higher values of the 'want parameter' for husbands, suggesting that they need more income than wives do, to attain the same level of welfare. However, looking at the overall values in the sample, the results indicate dissatisfaction with one's income because, in most cases, the real monthly income is considered bad. The MIQ highlighted the differences in attitudes between husband and wife and between areas. Husbands and wives had different ideas about the minimum income needed, suggesting that they differ in their appreciation of the level of household expenditures. We also found that with the same income, SP farmers were less satisfied than AP farmers. Looking at the determinants of the 'want parameters', the results show that variations in the individual welfare function are adequately explained by decision-making in the household, time allocated to crop activities and expenditures on private goods. Less time allocated by husbands to crops induces less welfare for women. Also, an increase in expenditures on clothing induces less happiness for men.

The Leyden approach was also used to assess the Leyden poverty line (LPL) and the subjective poverty line (SPL) in the two areas, indicating the income level associated with poverty. Both household size and household income explain husband's and wife's poverty levels. The estimated SPL and LPL show that men have higher needs than women, and that higher income needs are observed for better-educated people, for people with access to off-farm jobs, and for wives with access to credit.

Husbands and wives replied differently about their feelings towards happiness. While the husband's income and household size adequately predict his well-being, the wife's well-being is positively associated with her responsibility for managing finance and her education. Also, asking less permission gave her more happiness. Ethnicity matters in the estimation of well-being since Wolof and Peulh women are less happy than Sereer women. For men, well-being is not associated with ethnicity. Both men and women feel less happy in the AP area.

In our concluding Chapter 7, we make an inventory of our main findings and, subsequently, discuss the findings in view of their relevance and their place in the actual ongoing discussions on gender and household welfare and well-being. It appears that the complementary approaches used in analyzing productivity, gender roles, decision-making and welfare and well-being, are highly innovative in a developing country like Senegal. The results provide relevant knowledge for a better understanding of differences between husbands and wives in household consumption and welfare and well-being. In terms of policy implications, the results provide evidence that a better understanding of gender roles and husband-wife differences may improve policies towards the household and the community, and may avoid policy failures. For example, the choice of programs used for poverty alleviation should take into account the heterogeneity of spouses' preferences regarding the access to schooling, health treatment and care. Differences between areas showed significant variation that may help design specific policy actions.

Further research should go beyond the limitations of ours and may expand into an economic gender analysis of the agricultural chain (processing, distribution and marketing).

About the author

Fatimata Dia Sow, born in 1967 (Senegal) has a background in Sciences and Veterinary Medicine in EISMV (Ecole Inter-Etats des Sciences et Medecine Veterinaires) of Dakar, where she graduated in 1992. After her graduation as a Doctorate, she was working since 1993 as a researcher at ISRA (Senegalese Agricultural Research Institute). Mrs. Sow integrated teams of research and worked at household level for a better assessment of constrains and performance of production systems. In 1999, she was awarded a Master's degree in Louvain La Neuve at Belgium in Rural Economy where she obtained a Specialized Studies Diploma in Rural Economy in 2001 with distinction. Since then, she was working in socio-economic fields at ISRA covering the areas of integrated farming systems and gender issues. Between 2001 and 2004, she was working in a team as the supervisor of the socio-economic research of PROCORDEL in Senegal (Concerted project for research/development in livestock production in West Africa). Apart from the degrees, Mrs. Sow has also completed important specific courses in Data analysis and Statistics (Tuskeegee University: USA), in Development and Extension program (Munich, Israel), in Food security (Gembloux/Belgium) and Gender issues (Nairobi, Feldafing). Recently, she has been appointed to take part in the Training of Trainers in The Gender and Economic Policy Management Initiative-Africa aiming to accelerate achievement of the Millennium Development Goals. With respect to other responsabilities, Mrs. Sow was the Executive secretary of ASELFAE (African Women Leadership in Agriculture and Environment) between 2002-2004 and is membership of AWID (African Women in International Development) and the International Association for feminist Economist (IAFFE). In 2002, she was awarded PhD candidate at WU within the AWLAE program (African Women Leaders in Agriculture and Environment). This program aims at 'preparing a critical mass of self confident-pioneering women leaders to change the policies, programs and practices that affect women farmers in Africa'. Recent issue concerns strategies seeking to reverse the negative impact of HIV/AIDS on food security and poverty. In 2005, Mrs. Sow joined Wageningen University in the Economic of Consumers and Household Group.

AWLAE

African Women Leaders in Agriculture and the Environment

This thesis is one of a series of AWLAE thesis. It represents the fruits of collaboration between African Women Leaders in Agriculture and the Environment (AWLAE), Winrock International and Wageningen University and Research Centre. AWLAE is a pan-African Program that aims at training women professionals in the fields of agriculture and environment, to redress the existing gap between male and female representation in professions relating to the fields. AWLAE was initiated by Winrock International in 1989. Its headquarters are in Nairobi, Nairobi, Kenya.

Between AWLAE, WI and WUR a project formulated that was submitted for funding to the Minister for Development of Cooperation of the Netherlands Ministry of Foreign Affairs. The goal of the project was to build a cadre of well-trained African women professionals working in agriculture, environment and related sectors to enhance their academic standing and capacity to contribute to gender relevant research and policy-making on the role of women in food systems and the gendered impacts of HIV/AIDS on food security and rural livelihoods in sub-Saharan Africa. In April 2002 the project was granted. The Ministry agreed to fund twenty PhD Scholarships at Wageningen University and the additional leadership-in-change training for twenty women from eleven countries, ranging form East to West and Southern Africa. In June 2002 an agreement was signed between AWLAE, represented by its Regional Director, and the Director of the WUR Social Sciences Group, after which implementation of the project could start. The participating scholars were carefully selected from a large number of applications. The scholarships were widely advertised in relevant media in countries with AWLAE chapters, and the chapters concerned were actively involved in the recruitment and selection of candidates.

The following women participated in the AWLAE scholarship project:

Susana Akrofi	Ghana	Mariame Maiga	Ivory Coast
Namizate Binaté Fofana	Ivory Coast	Lydia Ndirangu	Kenya
Hirut Bekele	Ethiopia	Aifa Fatimata Ndoye Niane	Senegal
Joyce Challe	Tanzania	Faith Nguthi	Kenya
Fatima Dia Sow	Senegal	Carolyne Nombo	Tanzania
Stephanie Duku	Ghana	Regina Ntumngia Nchang	Cameroon
Rose Fagbemissi	Benin	Daisy Onyige	Nigeria
Kidist Gebreselassi	Ethiopia	Gaynor Paradza	Zimbabwe
Monica Karuhanga	Uganda	Corrie du Preez	South
Doris Kakuru	Uganda	Ekaete Udong	Nigeria

www.ingramcontent.com/pod-product-compliance
Lightning Source LLC
Chambersburg PA
CBHW061205240926
51846CB00006B/1121
* 9 7 8 9 0 8 6 8 6 1 5 8 3 *